Prevention of Actuator Emissions in the Oil and Gas Industry

Prevention of Actuator Emissions in the Oil and Gas Industry

Karan Sotoodeh

Gulf Professional Publishing is an imprint of Elsevier
50 Hampshire Street, 5th Floor, Cambridge, MA 02139, United States
The Boulevard, Langford Lane, Kidlington, Oxford, OX5 1GB, United Kingdom

Notices
Knowledge and best practice in this field are constantly changing. As new research and experience broaden
our understanding, changes in research methods, professional practices, or medical treatment may
become necessary.

Practitioners and researchers must always rely on their own experience and knowledge in evaluating and
using any information, methods, compounds, or experiments described herein. In using such
information or methods they should be mindful of their own safety and the safety of others,
including parties for whom they have a professional responsibility.

To the fullest extent of the law, neither the Publisher nor the authors, contributors, or editors, assume
any liability for any injury and/or damage to persons or property as a matter of products liability,
negligence or otherwise, or from any use or operation of any methods, products, instructions, or
ideas contained in the material herein.

Library of Congress Cataloging-in-Publication Data
A catalog record for this book is available from the Library of Congress

British Library Cataloguing-in-Publication Data
A catalogue record for this book is available from the British Library

ISBN: 978-0-323-91928-9

For information on all Gulf Professional publications
visit our website at https://www.elsevier.com/books-and-journals

Publisher: Joe Hayton
Senior Acquisitions Editor: Katie Hammon
Editorial Project Manager: Chris Hockaday
Production Project Manager: Surya Narayanan Jayachandran
Cover Designer: Alan Studholme

Typeset by SPi Global, India

Contents

Terms and definitions

A

AC voltage: Alternative current (AC) voltage constantly changes between positive and negative. The rate of these changes in direction is called frequency and is measured in hertz. AC is the opposite of DC voltage.

Accumulator: A storage vessel for pressurized hydraulic fluid that releases the fluid when it is required to maintain a steady-state flow of pump pressure and to dampen shock and vibration. Fig. 1.1 illustrates an accumulator for hydraulic oil, also called a hydraulic oil capsule.

Acme screw: A type of screw widely used in the oil and gas industry to convert rotary motion to linear movement. This type of screw can be used in an electrical actuator for valves with linear movement, such as through conduit gate valves.

Actuator: A mechanical or electrical device or component installed on a valve to automatically open and close the valve without any need for a manual operator. Actuators typically work with electricity, hydraulic fluid or air. The electric, hydraulic or pneumatic power is transferred to mechanical force to operate the valve.

Actuated valves: Automatic valves that are operated by actuators. Manual valves or manually operated valves are operated by a lever or handwheel and a hand force of an operator, and are the opposite of actuated valves.

API: The American Petroleum Institute (API) is the largest United States trade association for the oil and gas industry with headquarters in Washington, DC. Standards developed by the API are widely used in the oil and gas industry.

ASME: The American Society of Mechanical Engineers (ASME) is an American professional association that promotes the art, science and practice of multidisciplinary engineering and allied sciences around the globe.

ASTM: The American Society for Testing and Materials (ASTM) is an international organization that develops and publishes technical standards for a wide range of materials and products. ATSM standards are widely used to define the chemical and mechanical properties of piping and valve metallic components, the parameters of quality assurance tests and other requirements.

Austenitic stainless steel: A type of stainless steel that has relatively lower mechanical strength compared to other types, such as duplex and martensitic stainless steel. This type of stainless steel contains typically around 18% chrominum and 8% nickel.

Prevention of Actuator Emissions in the Oil and Gas Industry. https://doi.org/10.1016/B978-0-323-91928-9.00002-5

FIG. 1.1

An accumulator.

Axial valve: Axial valves, which are recommended for HIPPS application, are a type of on/off valve. The internal design of this valve is the same as that of axial flow check valves, but unlike axial check valves, it has a stem for valve operation.

B

Ball valve: A type of quarter turn valve used to start or stop fluid (on/off purpose). The shape of the valve closure member is like a ball with a hole inside. When the hole is parallel to the fluid flow, the valve is open and the fluid passes through the valve. The rotation of the closure member to 90 degrees closes the valve, as the hole in the ball then stands perpendicular to the flow direction and the solid part of the closure member stops the fluid.

Bar: A metric unit of pressure, but not part of the International System of Units (SI). One bar is defined as exactly equal to 100,000 pascal (Pa) (100 kilopascal (kPa)), or slightly less than the current average pressure at sea level (approximately 1.013 bar). Each bar is approximately 14.7 times 1 psi. Cf. psi.

Blowout: An uncontrolled release of crude oil and/or gas from a well. Modern wells have a blowout preventor as well as surface control subsea valves (SCSSVs) to prevent the blowout. Cf. SCSSV. Fig. 1.2 illustrates a catastrophic blowout from a well that set a platform on fire.

Butterfly valve: A quarter turn valve, like a ball valve, used for flow regulation and fluid isolation. The closure member of the valve is a disk that rotates 90 degrees between open and closed positions. Butterfly valves can be used instead of ball valves

FIG. 1.2

A blowout. *Courtesy: Shutterstock.*

as a less costly choice for fluid isolation in utility services such as water. Fig. 1.3 illustrates a very large 48″ butterfly valve with a rubber liner between its body and disk.

C

CAPEX: Capital expenditures (CAPEX) include expenses incurred by a company to buy technology, equipment, buildings, plants, etc. Cf. OPEX.

    Check valve: A nonreturn valve installed on a piping system to prevent backflow. The valve is opened by the fluid flow in the piping system, and closed by a spring or

Rubber liner

Handwheel and gear box

Disk

FIG. 1.3

48″ butterfly valve.

gravity force or both when the fluid moves back toward the upstream side of the valve. Check valves are thus opened and closed without any need for manual operation or actuators.

Choke valve: A choke is the preliminary control valve installed after the Christmas tree to reduce pressure and control the flow of the produced fluid from the wellhead. Choke valves are exposed to many operational problems, such as cavitation, erosion, etc.

Christmas tree: A component consisting of a combination of valves, piping and connectors installed on wellheads for different purposes, such as controlling the flow of the produced oil and gas from the well, or for the injection of chemicals, gas or water into the well. Christmas trees can be installed vertically or horizontally and may be installed on land or subsea. A Christmas tree installed on land is called a "dry tree" and one installed subsea is called a "wet tree." Fig. 1.4 illustrates a subsea Christmas tree.

Closed loop hydraulic system: A type of hydraulic system in which the hydraulic fluid is returned to the topside facilities through an umbilical system rather than being discharged to the sea. Cf. open loop hydraulic.

Closure member: Also called a valve obturator, a closure member is the part of a valve positioned inside the flow path to permit or prevent the flow. Different valves have obturators of diverse shapes and types, such as a ball, gate, disk or wedge.

Common return line: Please refer to return line.

Compressor: A facility or piece of equipment used to pressurize and move air or gas in a piping system. Compressors typically have two types of design: rotary or reciprocating.

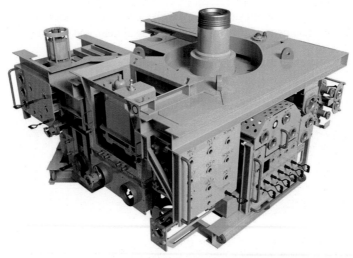

FIG. 1.4

Subsea Christmas tree.

Courtesy: FAVPNG.

Control panel: A component like a panel or a box that controls pneumatic and hydraulic actuators. The main aim of the control panel in an actuation system is to manage, command, direct and regulate the flow of air or hydraulic oil to the actuator. The air or hydraulic oil passes through the control panel before entering the actuator. Fig. 1.5 illustrates the control panel of a hydraulic actuator that includes tubing and different valves for control, such as solenoid valves, a filter and other components.

Control valve: A type of globe valve typically used for flow regulation. A control valve is normally equipped with a diaphragm actuator that works with air. A control valve regulates the flow by adjusting variables such as flow rate, pressure and temperature. Fig. 1.6 illustrates a control valve, in this case a globe valve, and a diaphragm actuator installed on a piping system.

Cycling: Cycling, also called valve cycling, means the action of both opening and closing. In fact, the valve is opened and closed, or closed and opened, in one operation cycle. Opening and closing the valve is more precisely called mechanical cycling.

D

DC voltage: Direct current refers to a constant and stable voltage circuit. DC is the opposite of AC voltage.

Direct drive: A mechanism in which the transmission of power and torque is performed directly by an electrical actuator without any interference of a gear. A gear is

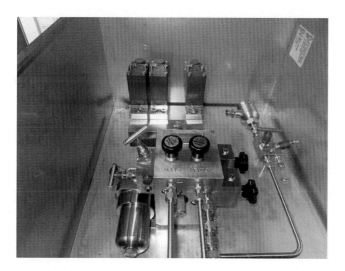

FIG. 1.5

Control panel for a hydraulic actuator.

FIG. 1.6

A control valve installed on a piping system.

Courtesy: Shutterstock.

typically required if the motor cannot generate enough torque or force to move a mechanical component such as an industrial valve.

Directional control valve (DCV): DCVs are the part of a hydraulic control system that allows the flow of fluid to move to different paths and locations.

Disk: The closure member of a through conduit gate valve is called a disk. The closure member of a valve is defined as a valve internal that can close or open the valve to stop the flow or allow the flow passage, respectively.

Double acting actuator: A springless actuator that gets both opened and closed by means of air or hydraulic pressure. Double acting actuators are categorized as fail-as-is actuators that stay in their last position upon losing power. The term 'double acting' is not applicable to electrical actuators.

E

Elastomer: A type of polymer with high viscosity and elasticity. Rubber materials used in valves, actuators and compensation systems are elastomers.

Electrical actuators: This type of actuator is powered by electricity and contains a motor and some gear boxes to increase its torque output. Electrical actuators are considered an environmentally friendly design.

Electro-hydraulic actuators: A type of hydraulic actuator that works with a motor, thus eliminating the need for separate hydraulic pumps and tubing. Electro-hydraulic actuators are simpler and more reliable than hydraulic actuators. Fig. 1.7 illustrates an electro-hydraulic actuator.

Emergency shut down (ESD) system: An ESD system is part of plant safeguarding. Its purpose is to prevent the escalation of abnormal conditions into major hazardous events.

Emergency shut down (ESD) valve: An actuated valve designed and selected to stop the fluid upon detection of a dangerous event. ESD valves should be capable of fast closing. Emergency shut down valves should have a fail safe mode of operation.

F

Factory acceptance test (FAT): A process of testing equipment or components, such as valves and actuators, performed after assembly to verify that the components or facilities are made according to the relevant standards and specifications. A pressure test is considered to be part of FAT.

Fail-as-is actuator: An actuator that returns to its latest position in the event of power loss. Electrical actuators in the non-subsea sector of the oil and gas industry are mainly categorized as fail-as-is actuators. Double acting actuators that are normally powered by oil or air are fail-as-is actuators, since upon losing their source of power, they stay in their last position.

Electrical actuator (motor)

Hydraulic actuator

FIG. 1.7

An electro-hydraulic actuator.

Fail safe actuator: An actuator that returns to its home position, which is either fully open or closed, in the event of losing its source of power supply, which may be air or hydraulic fluid. If the actuators return to closed position in the event of losing power supply, they are known as fail safe closed (FSC) actuators. The actuators are called fail safe open (FSO) if they return to open position in the event of losing power. Fail safe actuators are also called "spring return actuators," since they are equipped with a spring that returns the actuator to a fail safe mode.

Fieldbus: A kind of technology that is increasingly used for data transmission in process automation applications. Cf. Profibus.

Flange: A ring categorized as a bulk piping component used to join two pieces of pipe together or to connect piping to equipment in general. Flanges are typically welded to the connected pipe, but other types of flange connections exist, such as threaded. Two flanges are connected with bolting (bolts and nuts) and a gasket is placed between two pieces of flanges for sealing.

Flare: A system that includes piping and a flare stack in which the produced hydrocarbon, oil and mainly gas, is burnt. Typically, flaring is performed to release overpressure gases and liquids to the environment.

Flare stack: A flare stack is a part of a flare system in an industrial plant where all overpressurized gases and liquids are released by a pressure safety valve and burnt to prevent overpressurizing some of the equipment and facilities in a plant.

Flowline: A flowline, also called a pipeline, is defined as a segment of piping that transfers the fluid coming out from the wellhead to the topside facilities from the manifolds. A flow line is not typically connected directly to the host facility. Another pipe, called a riser, connects the host facility and the flowline, as illustrated in Fig. 1.8.

Flying lead: Flying leads provide electrical, hydraulic and chemical connections from an umbilical termination assembly (UTA) to a subsea distribution unit (SDU), and further from the SDU to the tress and manifolds. Flying leads can be hydraulic

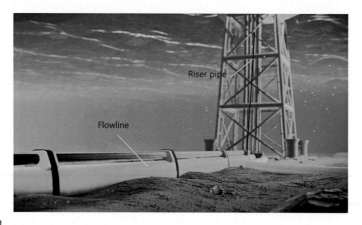

FIG. 1.8

A subsea flowline and riser connection to a platform.

Courtesy: Shutterstock.

flying leads (HFL) or electrical flying leads (EFL), which are used for the transmission of hydraulic and electrical power, respectively.

FPSO: Floating Production, Storage and Offloading (FPSO) ships are used in the offshore sector of the oil and gas industry for the production and processing of oil and gas. FPSO units are used for subsea oilfield development in deep areas, such as depths of 1 or 2 km, where the installation of a platform is not practical or feasible. Fig. 1.9 illustrates an FPSO for the development of an oilfield.

Fugitive emission: An unintentional and undesirable emission, leakage or discharge of gases or vapors from pressure containing equipment or facilities, or from components inside a plant such as valves, piping flanges, pumps, storage tanks, valves, compressors, etc. Fugitive emission is also known as leak or leakage.

G

Gate valve: An on-off valve that works by inserting a rectangular gate or wedge into the flow of the fluid. The fugitive emission standard for gate valves is covered by API 624 and ISO 15848-1. It should be noted that the stem motion in a gate valve is typically linear, which creates a lot of friction between the valve stem and packing. This friction can cause packing wear and tear as well as leakage. Gate valves are available in different types, such as slab, expanding and wedge. Wedge gate valves have a sealing element in the shape of a wedge. A wedge gate valve is a torque seated valve, meaning that the wedge is expanded from both sides due to the stem force and provides sealing. The expansion of the valve closure member due to the stem axial force is called "wedging effect." Expanding gate valves are also torque seated valves with

FIG. 1.9

An FPSO.

Courtesy: Shutterstock.

a closure member in two sections, one male and the other female. Slab gate valves have a flat disk or closure member that provides sealing due to the fluid pressure. Slab gate valves, unlike expanding and wedge gate valves, are not torque seated.

Gear box: Gear boxes are typically used with handwheels to facilitate valve operation. This is a simple and cheap method of valve operation in which gears are used to increase the force and efficiency produced by the operator moving the handwheel. The gears inside the gear box are wheels with teeth that slot together. Let us suppose that a gear box contains two gears, as illustrated in Fig. 1.10; the first one on the left, the "driving" gear, is smaller and has 20 teeth. The second, "driven" gear is larger with 40 teeth. The gear box in this case increases the input force applied by the operator through the handwheel, since the driving gear has lower numbers of teeth. The amount of increase depends on the gear ratio, which is calculated through Formula (1.1).

Gear ratio calculation

$$\text{Gear ratio} = \frac{\text{Number of teeth on driven gear}}{\text{Number of teeth on driver gear}} \tag{1.1}$$

Thus the gear ratio in this case is equal to two. It means that if the operator input force on a handwheel with two hands is 300 newton (N), the gear box in the example given above will increase the force to 600 N.

Global warming: Global warming is the gradual heating of the earth's surface, oceans and atmosphere caused by air pollutants such as methane and carbon dioxide and the greenhouse effect.

Globe valve: A type of valve used for flow regulation or throttling. As illustrated in Fig. 1.11, the fluid makes two 90-degree turns inside the valve, which creates a significant pressure drop.

Greenhouse effect: The greenhouse effect is a natural process that warms the earth's surface. When the sun's energy reaches the earth's atmosphere, some of the heat is reflected back to space and the rest is absorbed and re-radiated by greenhouse gases.

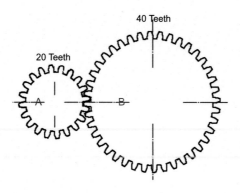

FIG. 1.10

Gears in a gear box for increasing input force.

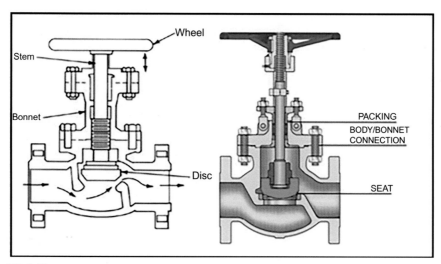

FIG. 1.11

Globe valve.

H

Handwheel: A wheel installed on some manually operated valves for operating (opening and closing) the valves. The handwheel design for valves in the oil and gas industry is a spoke and rim design for better gripping, as illustrated in Fig. 1.12.

Hardware fault tolerance: The ability of a system or component to maintain the required safety instrumented function (SIF) in the presence of zero, one or more dangerous faults in the system.

FIG. 1.12

Spoke and rim handwheel.

Hazardous air pollutant (HAP): HAPs, also referred to as toxic air pollutants, are those that can cause serious health problems such as cancer. The EPA has identified 187 types of hazardous air pollutants, such as lead, ozone, nitrogen dioxide, etc.

Hazardous area: A hazardous area is defined as any location where there is a risk of explosion. Typically, three different hazardous zones are identified for vapors, mists and flammable gases.

High integrity pressure protection system (HIPPS): A type of safety instrument system designed to prevent overpressurization of the plant. HIPPS valves are used to shut down the source of high pressure before it exceeds the design pressure of the system. HIPPS valves, like emergency shut down valves, should have a high speed of closure during operation.

Hooke's law: A law in physics that states that the force (F) required to extend or compress a spring by a distance (x) can be calculated through the formula $F = Kx$, where K is the contestant factor of the spring, which depends on the spring characteristic.

HSE: Health, safety and environment (HSE) is a methodology that implements practical aspects of protecting the environment and ensuring occupational health and safety. HSE implementation is a concern of many companies, especially those in the oil and gas industry, to provide guidelines and procedures from the high and general level down to the detailed design level to prevent HSE problems such as oil spillage, fugitive emission and leakage, etc.

Hydraulic head pressure: The head of pressure resulting from a column of hydraulic fluid.

Hydraulic power unit (HPU): Sometimes called a hydraulic power pack, an HPU is a system that contains a motor, a fluid reservoir and a pump. An HPU stores hydraulic fluid and transports it to the required destination. An HPU to transmit hydraulic fluid to a subsea system is normally installed on a host facility, such as a platform or ship.

Hydraulic return line: Please refer to return line.

Hydrate: Ice-like, solid particles that can be formed from vapor or water inside a gas when the gas is subjected to low temperature and high pressure. The possibility of hydrate formation due to the above-mentioned conditions occurs under the ocean floor. Mono-ethylene glycol (MEG) and methanol can be injected into the gas to prevent hydrate formation.

I

IEC: The International Electrotechnical Commission (IEC) is an international standard-setting body for electrical, electronic and related technologies collectively known as "electrotechnology."

ISO: The International Organization for Standardization (ISO) is a standard-setting body whose standards are often referred to in the oil and gas industry.

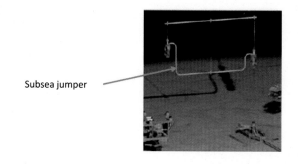

Subsea jumper

FIG. 1.13

Subsea jumper.

J

Jumper: A short pipe piece connector widely used in subsea oil and gas production systems to connect two subsea structures, such as a Christmas tree to a Christmas tree or a Christmas tree to a manifold. Fig. 1.13 illustrates a subsea jumper.

N

Nitrile rubber: Also called Buna N or NBR, nitrile rubber is a type of synthetic rubber used for sealing in valves and actuators. It can also be used in closed compensation systems.

L

Lever: A lever is typically used for opening and closing quarter turn valves. The lever can indicate the position of the valve; if it is parallel to the pipe, the valve is open, and if it is perpendicular to the pipe, the valve is closed. Lever operation is typically used for small and/or low-pressure class valves that require less force or torque for operation. Fig. 1.14 illustrates a lever-operated ball valve being tested on a test bench. The lever direction indicates that the valve is in half open position.

Linear actuator: A linear actuator is an actuator that creates motion in a straight line, unlike scotch yoke or rack and pinion actuators. This type of actuator is mainly used for gate valves.

Linear motion valve: A valve whose stem is characterized by linear movement and whose closure member moves upward and downward to open and close the valve. Valves with linear motion or linear stem motion are equipped with a linear actuator. Gate and globe valves are linear motion valves. Fig. 1.15 illustrates a gate valve in fully open and closed positions. The stem and gate move upward to keep the valve in fully open position, as illustrated in the right side of Fig. 1.15. On the left side of the figure, the valve stem and disk are moving downward to keep the valve in closed position.

FIG. 1.14

Half open lever-operated ball valve.

Courtesy: Elsevier.

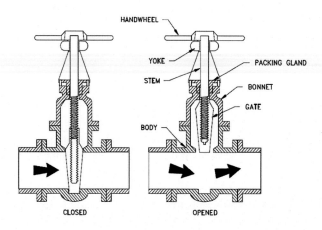

FIG. 1.15

Linear motion of a gate valve stem and closure member in fully opened and closed positions.

M

Manifold: Manifolds in the subsea sector of the oil and gas industry are made of structures, piping and valve components that are used to gather fluid from different wells and combine them into one header. Therefore manifolds are used to integrate, distribute and control the flow. Manifolds are used to simplify the subsea system by

FIG. 1.16

Subsea manifold.

Courtesy: Shutterstock.

minimizing the number of subsea pipelines and risers. Fig. 1.16 illustrates a subsea manifold submerged in water to be installed subsea.

Manually operated valves: "Manual valves" are defined as those operated by an operator (human) by turning a lever or handwheel between open and closed positions. Actuated valves are the opposite of manual valves; the former are operated automatically through actuator force.

MAST (maximum allowable stem torque): The maximum torque/thrust that can be applied to a valve train without risk of damage, as defined by the valve manufacturer/supplier.

Methane: Methane, with formula of CH_4, is the lightest and simplest form of hydrocarbon. Methane is a powerful greenhouse gas. It is flammable and is used for fuel worldwide. Methane is a colorless gas that has no smell at low concentration. However, it has a sweet smell at high concentration. The mixture of methane in air with a concentration above five percent could be explosive. Methane is considered a volatile organic compound (VOC).

Methanol: A type of alcohol used as a chemical in the oil and gas industry. Methanol is injected into gas services in the piping system to prevent hydrate formation. The application of methanol in the oil and gas industry is the same as mono-ethylene glycol (MEG).

Mono-ethylene glycol (MEG): A type of chemical widely used in the oil and gas industry to prevent hydrate formation.

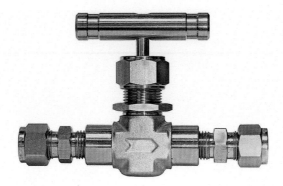

FIG. 1.17

Needle valve. *Courtesy: Shutterstock.*

N

Needle valve: A type of valve used for flow regulation (see Fig. 1.17). The operation principle of a needle valve is similar to that of a globe valve with two main differences; the first is that the type of plug or closure member is like a needle and the second is that the valve is small and is used for a small amount of flow regulation.

NORSOK: A set of standards developed by the Norwegian petroleum industry to ensure adequate safety, add value and improve cost effectiveness for petroleum industry developments and operations.

O

Offshore oil and gas: Oil and gas industry operations that take place off the shore and in the sea. Some of these activities are conducted underwater; this sector is known as subsea. Other activities are performed on platforms or ships located in the water.

Open loop hydraulic system: A hydraulic system in which the hydraulic fluid is discharged into the marine environment during the actuation of the subsea valves. A closed loop hydraulic system is the alternative solution.

Operator: Personnel/human being who operates a valve.

OPEX: Operational cost.

P

Partial stroke test: A type of test performed on emergency shut down and HIPPS actuated valves in order to detect possible dangerous failures and improve the safety and reliability of these components. A valve undergoing this test is partially opened and closed, which is an advantage since it does not interfere with production.

Pipeline end manifold (PLEM): A PLEM is a subsea structure (a simple manifold) placed at the end of a pipeline to connect a rigid pipeline with another structure or additional piping.

Pipeline end terminal (PLET): A PLET is a simple structure used to connect two pieces of pipe together, such as a flowline to a jumper.

Piston-crank mechanism: A mechanism in rotary diaphragm actuators that converts the linear movement of the piston to rotary movement generated by a crank. The piston is connected through a rod to the crank.

Pitting corrosion: A local type of corrosion in which cavities and holes are created inside the material. Pitting corrosion is caused by a corrosive environment that contains chloride.

Pitting resistance equivalent number (PREN): A number that is calculated based on the elements such as chloride, molybdenum and nitrogen to assess the resistance of the material against pitting corrosion. PREN is calculated through Formula (1.2).

PREN calculation

$$PREN : Cr + 3.3 Mo + 16 N \qquad (1.2)$$

where Cr: chromium, Mo: molybdenum, N: nitrogen

Pressure containing components: Sections of the valve in which failure to function leads to the leakage of the internal fluid to the environment. The body, bonnet, stem and bolting are categorized as the pressure containing components of valves.

Pressure relief valves: A relief valve or pressure relief valve is a type of safety valve used to control or limit the pressure in a system; the pressure might otherwise build up and create a process upset, instrument or equipment failure or fire. Fig. 1.18

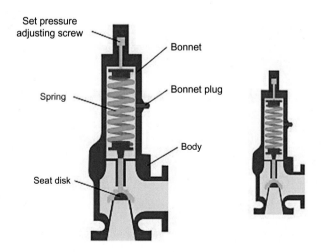

CONVENTIONAL SPRING LOADED
PRESSURE RELIEF VALVE

FIG. 1.18

Pressure relief valve. *Courtesy: Shutterstock.*

illustrates a typical pressure relief valve. The overpressure fluid enters the valve from the inlet port, pushes the disk upward by overcoming the spring force, and exits the valve through the outlet port. Pressure relief valves, unlike gate, globe and ball valves, do not have any stem, stem seal or stem packing.

Pressure test: Components that work under pressure, such as industrial valves and pneumatic and hydraulic actuators, should undergo a pressure test to check the mechanical or structural integrity of the product and ensure that no leakage can escape from the product to the environment. Various pressure tests are performed to make sure that the internal leakage inside the product is zero or within the allowable limit. A pressure test is considered to be part of the factory acceptance test (FAT). Fig. 1.19 illustrates the pressure test of a manual ball valve.

Process flow diagram (PFD): A PFD is a diagram commonly used in the oil and gas industry and generated by the process department to show the main process, including the main piping and equipment. A process flow diagram shows the process of oil, gas and water treatment in oil field developments. PFDs in refineries and petrochemical plants illustrate how the feed fluid service that enters the plant is converted to the final product.

Process shut down system: Part of a plant safeguarding system designed to minimize the frequency and occurrence of a facility operation outside the defined operation envelope. Valves and actuators connected to a process shut down system should have a fail safe mode of operation.

Profibus: Profibus technology provides automated communication, and is a standard for fieldbus. Profibus and fieldbus are defined as industrial communication

FIG. 1.19

Pressure test of a manual ball valve.

systems that use a range of components, such as copper cable, fiber optics or wireless signals to transfer data between field devices, such as sensors and actuators, to the control room or management system.

Psi: Pounds per square inch (psi) is the amount of force, in pounds, applied on a square inch.

Q

Qualification: Newly designed subsea valves and actuators should undergo a series of tests known as qualification tests.

Quarter turn: Valves whose stem and closure member rotates just 90 degrees between opening and closing positions. Ball and butterfly valves are categorized as quarter turn valves.

Quick exhaust valve: Quick exhaust valves are used in control panels in order to achieve rapid piston speed on the return stroke of single and double acting pneumatic actuators. Quick exhaust valves are not used on hydraulic actuator control panels.

R

Rack and pinion actuators: A type of rotary actuator that can provide rotary motion for ball valves with a 90-degree rotation of the closure member. Two gears are used in rack and pinion actuators; the first is a linear gear called a rack, and the second is called a pinion. The rack engages teeth in a circular gear called a pinion.

Return line: Also called a "hydraulic return line" or "common return line." A line that transports the hydraulic fluid back to topside host facilities after being used in a subsea hydraulic actuator. Return lines are part of a closed loop hydraulic distribution system.

Rotating stem valve: A valve that is opened and closed by a 90-degree rotation of the stem, such as a ball valve's stem movement.

Rotor: Part of an electrical motor, a rotor is a rotating electrical component that consists of a variety of electro-magnets arranged around a cylinder, with the poles facing toward the stator poles. Cf. stator.

ROV: A remotely operated vehicle (ROV) or remotely operated underwater vehicle is a machine used to explore the ocean and, more importantly, perform underwater operations. Subsea actuators are typically equipped with ROV override so that they are operable by an ROV independently from a source of power.

ROV bucket: A bucket installed on subsea valves and actuators for manual operation. Typically, a torque tool is inserted into the bucket by an ROV for manual operation.

S

Safety analysis report (SAR): A report prepared by valve and actuator suppliers to prove that the valves, actuators and associated control system can provide the required safety integrity level (SIL).

Safety factor: This term is used in actuator sizing; it is defined as the ratio of the actuator torque value to the valve torque value, as per Formula 1.3.

Safety failure fraction (SFF): The ratio of the average probability of safe failure plus dangerous detected failure to the total probability of the failures, including safe and dangerous failures. There is a correlation between SFF and SIL, refer to IEC 61508 standard.

Safety instrumented system (SIS): A system including both software and hardware responsible for safe operation within the designed limits. Fig. 1.20 illustrates a safety instrumented system, including a sensor, logic solver and final element, which is an actuated valve in this case. As an example, the sensor measures the pressure and sends a signal to a logic solver. The logic solver compares the measured pressure from the sensor to the desired pressure range. If the measured pressure is higher than the maximum design temperature, the logic solver sends a signal to shut down the actuated valve.

Safety integrity level (SIL): Safety integrity level (SIL) is a part of an international standard such as IEC 61508 that provides suppliers and end users with a common framework to design products and systems for safety-related applications. SIL provides a scientific and numeric approach to specifying and designing safety systems, enabling the risk of failure to be quantified. There are four levels of SIL: 1, 2, 3 and 4. A higher SIL means higher safety and less possibility of failure.

Scotch yoke actuator: A type of rotary actuator suitable for quarter turn valves, such as ball and butterfly valves. This type of actuator can work with hydraulic fluid or air and could be either single acting or double acting. A scotch yoke actuator has a shaft or piston, and the transformation of the linear movement of the shaft to rotary motion is performed through the yoke mechanism. Fig. 1.21 illustrates a spring return scotch yoke actuator.

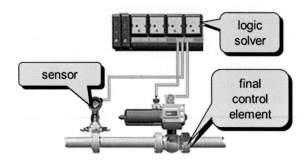

FIG. 1.20

A safety instrumented system (SIS).

Yoke   Spring

FIG. 1.21

Scotch yoke spring return actuator.

Seawater head: The head of pressure resulting from a column of water. As a rule of thumb, every 10 m of water column produces 1 bar of pressure.

Separator: A pressure vessel used in the oil and gas industry to separate the two phases of oil and gas, or the three phases of oil, gas and water from each other.

Shut down valve: A normally open valve that closes in the event of failure in the system. As an example, a shut down valve could be used upstream or before a separator; it would close if an overpressure scenario occurred inside the separator.

Spring return actuator: A failsafe actuator that returns to its home position via spring force upon losing power. Spring return actuators can be failsafe open or closed. As an example, a failsafe closed hydraulic actuator is opened by hydraulic force and closed by spring force upon losing hydraulic force. Spring return actuators are also called single acting actuators (see Fig. 1.22).

Solenoid valve: An electromechanically operated valve used on the control panel of both pneumatic and hydraulic actuators. This type of valve opens and closes by the start and stop of electricity into the valve, respectively.

Stainless steel: A group of iron-based alloys that contains a minimum of approximately 11% chromium. Stainless steel grade 316 is a type of austenitic stainless steel that has approximately 18% nickel, 8% chromium and 2% molybdenum. Austenitic is a category of stainless steel that has less mechanical strength compared to other types.

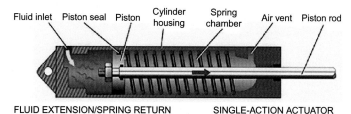

Fluid inlet Piston seal Piston   Cylinder   Spring   Air vent Piston rod
                                   housing   chamber

FLUID EXTENSION/SPRING RETURN   SINGLE-ACTION ACTUATOR

FIG. 1.22

Spring return/single acting hydraulic actuator.

Courtesy: Springer.

Stator: The stationary part of an electrical alternative current (AC) motor. It is made of steel and contains some wiring, such that when the three-phase electricity passes through the wiring, the stator produces a rotating magnetic field that causes the rotor to rotate.

Stem: A stem or shaft is the part of a valve that transfers the load from the valve operator device, which could be a gear box or an automatic component (actuator), to the valve internals. Valve stems can have rotary or linear motion.

Subsea control module (SCM): A part of a subsea control system used for monitoring and controlling subsea facilities and assets.

Subsea distribution unit (SDU): A subsea structure used to distribute hydraulic fluid and chemicals to the required subsea facilities and components. Hydraulic fluid is used in subsea actuators, and chemicals are injected inside the piping system for different purposes such as wax and scale prevention, etc.

Subsea oil and gas industry: A sector of the offshore oil and gas industry in which the facilities, such as piping, valves, pumps, etc., are installed underwater. The depth of subsea facility installation could be shallow, e.g. less than 200 m, or as deep as 3 km.

Subsea pressure compensation system: A storage tank or capsule with hydraulic oil inside to eliminate the effect of the seawater head at the installed depth. A compensation system is typically installed on actuators, valve gear boxes and remote-operated vehicles (ROVs). The pressure of oil inside the compensation system is almost equal to the seawater head pressure at the depth of equipment installation.

Surface controlled subsurface safety valve (SCSSV): A subsurface safety valve (SSSV) is a type of emergency shut down (ESD) valve installed inside a well to isolate or shut down the well in case of any hazardous event such as a blowout. SSSVs that are controlled and operated from surface facilities are called SCSSVs.

T

Through conduit gate valve: A type of on/off valve used for flow isolation, especially in process fluid such as oil and gas. There are two types of through conduit gate valves: slab and expanding. Through conduit gate valves are widely used on subsea Christmas trees and manifolds.

Topside oil and gas: The sector of the offshore oil and gas industry where the facilities and equipment are installed above the water on a platform or ship such as an FPSO.

Torque: The function of force and the distance necessary for opening and closing a valve. The torque produced by an actuator is typically higher than the torque required for the operation of a valve. The ratio of actuator torque to valve torque is defined as the safety factor, as per Formula 1.3.

Actuator safety factor calculation

$$\text{Safety factor} = \frac{\text{Actuator torque}}{\text{Valve torque}} \qquad (1.3)$$

Torque to weight ratio: An indication of actuator performance calculated through the ratio of torque to the weight of the moving part of an actuator. Electrical actuators have a high torque to weight ratio, but pneumatic actuators have higher torque to weight ratio compared to electrical actuators.

U

Umbilical: A component used in the subsea sector of the oil and gas industry that includes hoses and cables in order to transfer electricity and hydraulic power from topside to subsea equipment and components. Fig. 1.23 illustrates a subsea umbilical.

Umbilical termination assembly (UTA): A UTA, also called a subsea umbilical termination assembly (SUTA), is a subsea structure where all the umbilicals end. Electrical and hydraulic connections, called flying leads, connect the SUTA to the subsea distribution unit (SDU).

FIG. 1.23

Subsea umbilical.

V

Valve: A mechanical component used in a piping system for safety purposes, and/or to open and close the fluid passage, prevent fluid backflow and control the flow.

Valve stroke: The movement of a valve between fully closed and fully open, or from fully open to fully closed.

Volatile organic compound (VOC): VOCs are organic compounds or chemicals that have high vapor pressure at ordinary room temperature and low water stability. Many VOCs are synthetic chemicals used in the different industries. VOCs are typically emitted as gases from solids or liquids.

W

Wear: The interaction of the surface of a material with the environment or another material that leads to the removal of some metal from the material surface.

Wellhead: A component that can be installed on the land, or in the sea or ocean as a barrier between the drilled well and the production equipment. In addition, a wellhead provides structural and pressure integrity between the components inside a drilled well, such as the casing and tubing hanger. Fig. 1.24 illustrates a subsea wellhead.

Wing valve: The part of a Christmas tree used to shut down the flow from the producing well. Typically, Christmas trees have two wing valves, as per Fig. 1.25; one is a preliminary wing valve that has the function stated above, the second is normally used for injecting chemicals into the well.

FIG. 1.24

Subsea wellhead.

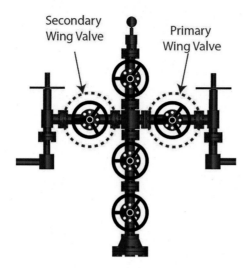

FIG. 1.25

Wellhead with two wing valves.

Introduction to actuators

2.1 Actuator definition

Every valve requires a means of operation. In terms of operation, all the valves except for pressure relief and check valves can be divided into two groups; manual or actuator operation. Manually operated valves or manual valves are those that are operated through a lever or a handwheel plus a gear box. Fig. 2.1 illustrates a manually operated valve with a handwheel and a gear box that is under operation by an operator (human).

An actuator is a mechanical or an electrical component installed on the top of the valve (see Fig. 2.2) for automatically moving and controlling the opening and closing of the valve. Thus actuators are among the most important elements in valve automation. Actuators can automate the valves in such a way that no human interaction is required for operation of the valves. Usage of the actuators is more common nowadays compared to that in the past, for a variety of reasons, such as less need for personnel to operate the valves, precise control of the valve operation, ease, safety, speed, and reliability of the valve operation. As an example, valves that require frequent operation or those that are located in remote and hazardous locations should be actuated. Valve actuators have several functions, such as moving the closure member of the valve to suitable open or closed positions, holding the valve closure member in the desired position, and providing sufficient force or torque for shutting down the valve in an accepted or allowable leakage class.

The working principle of the actuators is by converting an external energy power, such as air, hydraulic power, or electricity, into a mechanical motion. The quality of the valves is largely dependent on the proper metallurgy and material selection of the valve components, on the mechanical strength of components, especially with the pressure-containing parts, and on proper machining and tight tolerances, etc. However, the performance of the valve, such as the ability of the valve to open and close, which is also called cycling, is largely dependent on the actuators. Actuators are found and used in a variety of processing plants, such as refineries, petrochemical plants, pipelines, nuclear processers, etc. Actuators can work with different sources of power, such as air, hydraulic oil, electricity, or a combination of hydraulic power and electricity, which are discussed in the following section.

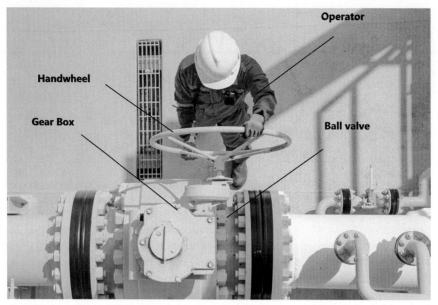

FIG. 2.1

A manual valve in the plant under operation by an operator.

Courtesy: Shutterstock.

FIG. 2.2

Actuated valves.

2.2 Actuators' source of power

2.2.1 Nonsubsea oil and gas

Nonsubsea oil and gas covers all different oil and gas units that are not subsea, e.g., refineries, petrochemical plants, and topside offshore. Actuators in these sectors use three main power sources to move the industrial valves: air, hydraulic oil, and electricity, all of which are briefly explained in this section. It is important to consider the combination of electrical and hydraulic power sources for moving the actuator.

2.2.1.1 Air

Compressed and pressurized air with the pressure value of around 5–9 bar is one of the main power sources of the actuators. Actuators that are working with air are called pneumatic actuators. Pneumatic actuators could be used for both linear and quarter turn (rotary) valves. The main advantage of pneumatic actuators is that they are working with air, which is a safe and environmentally friendly fluid, unlike hydraulic oil. The other advantages of pneumatic actuators are fast operation and the possibility of keeping the valve either in the closed or open position upon losing the air supply, for a failsafe feature; these are also called spring return actuators. Pneumatic actuators have a control panel or system that is more compact than the control panel or system in hydraulic actuators (see Fig. 2.3 for control panel comparison between pneumatic and hydraulic actuators). Pneumatic actuators have a higher

FIG. 2.3

Control panel/system of pneumatic actuator (left side) vs hydraulic actuator (right side).

Courtesy: Elsevier.

torque-to-weight ratio compared to electrical actuators. The hydraulic actuators contain fewer moving parts, so they are exposed to less wear and are easier to repair.

Two disadvantages of pneumatic actuators are relatively high cost compared to the electrical actuators, as well as less precise actuation due to compressibility of the air during the actuator operation. The limitation of pneumatic actuators is that they cannot provide sufficient torque or force for operation of the large-size and high-pressure class valves. Thus pneumatic actuators should be changed to hydraulic actuators for large-size and high-pressure classes. There is no clear split between the pneumatic and hydraulic actuators selection regarding the size and pressure class. Fig. 2.4 illustrates a 24″ slab gate valve in pressure class 1500 equal to 250 bar pressure as per ASME B16.34 standard for valves, which is going to be installed on a topside platform in the offshore oil and gas industry. The slab gate valve is equipped with a hydraulic actuator, since using a pneumatic actuator with the same size as the red color hydraulic actuator in the figure cannot produce enough force for the valve operation. In fact, a pneumatic actuator that would be able to operate a slab gate valve, as illustrated in Fig. 2.4, should be about 20 to 40 times bigger than the hydraulic actuator in the figure. Fabrication and installation of such a huge pneumatic actuator for the illustrated valve is not practical, so a hydraulic actuator is selected for the valve.

Leakage of the compressed air from the actuator does not harm the environment but has other negative consequences, such as increasing the cost of operation as well as killing or damaging the human. In fact, losing the compressed air in the plant is

FIG. 2.4

A 24″ CL2500 slab gate valve with a hydraulic actuator for offshore oil and gas industry.

Courtesy: Elsevier.

costly. Leaks of compressed air can waste thousands of dollars of electricity per year. The losing of compressed air leads to losing the pressure and jeopardizing the connected product, an actuator performance. As an example, losing air pressure supplied to the actuators causes insufficient force for the valve operation connected to the actuator, so the valve cannot be fully opened or closed. The amount of air leakage in an oil and gas unit could be estimated as high as 30% of the air flow outputs from the compressors inside the unit. However, not all of the air loss or leakage are associated with pneumatic actuators and relevant systems. In addition, OSHA, which is known as the Occupational Safety and the Health Administration, says that air pressure above 30 psi, which is equal to almost 2 bar, is dangerous for a human. Inhaling compressed air can damage the lungs, and compressed air entering the ear can damage the ear and brain. Pneumatic actuators are not used in the subsea oil and gas industry.

2.2.1.2 Hydraulic oil

Hydraulic actuators, as with the pneumatic types, covert fluid pressure into mechanical movement. The working principle of a hydraulic actuator is the same as that of a pneumatic actuator, with the difference being the source of fluid power. Instead of pneumatic air, pressurized hydraulic oil, with a pressure of around 160–200 bar, is used for the operation of hydraulic actuators. Thus hydraulic actuators can produce a much higher force or torque for a valve operation, compared to the pneumatic actuators. In fact, hydraulic actuators are a more suitable choice for valves that require a high amount of torque or force for valve operation. Large-size valves in high pressure classes typically require a higher amount of torque. The type of valve, in addition to the size and pressure class, is another parameter that affects the required force or torque for the valve operation. As an example, ball valves have heavier internals, such as closure members, compared to that of a butterfly valve with the same size and pressure class. Therefore the required torque for the operation of a ball valve is typically higher than that of a butterfly valve with the same size and pressure class. Fig. 2.5 illustrates a 38″ CL1500 ball valve with a hydraulic actuator on the top in the red color.

Hydraulic actuators have the following advantages:

- Hydraulic actuators are more compact than pneumatic actuators, since the oil pressure is higher than air pressure. In fact, since the oil pressure in a hydraulic actuator is higher than the air pressure in a pneumatic actuator, less volume of oil compared to air is required to create the same amount of force.
- Hydraulic actuators are more precise than pneumatic actuators, since the oil is not compressible.
- Hydraulic actuators provide very high-speed operation for large-size and high-pressure class valves. The speed of hydraulic actuators is higher than that of both pneumatic and electrical actuators.
- Hydraulic actuators have the highest torque-to-weight ratio in comparison with that of pneumatic and electrical actuators.
- The hydraulic actuators contain fewer moving parts, so they are exposed to less wear and are easier to repair.

FIG. 2.5

38" CL1500 ball valve with hydraulic actuator.

Courtesy: Elsevier.

The disadvantages of a hydraulic actuator are summarized as follows:

- The high pressure of hydraulic fluid handling requires a high level of safety and precaution.
- Leakage of the hydraulic oil, unlike with air, has negative impacts on the environment.
- The control system of a hydraulic actuator is typically a large box, as illustrated on the right-hand-side picture of Fig. 2.3, that requires a large space. Having a large space in the offshore oil and gas industry is always a challenge.

2.2.1.3 Electricity

Electrical actuators use electricity or electrical power for operation of the valves. The electrical motor may work with different AC or DC voltage. It is common to use three-phase AC motors as a driving force. However, single-phase AC or DC voltage motors do exist as alternatives. Electrical actuators are also called motors, which are supplied with a gear box if there is a requirement for a higher torque or force value for the valve operation. As an example, just an electrical motor may be sufficient for operation of a 4" butterfly valve in pressure class 150, but the motor should be upgraded with a gear box for a butterfly valve of 30" size and a pressure class of 150, to provide higher torque for the valve operation. The advantages of electrical actuators are summarized as follows:

- Electrical power is relatively cheap, easy to handle, and available in different plants.

- Electrical actuators have a high torque-to-weight ratio.

0% opening

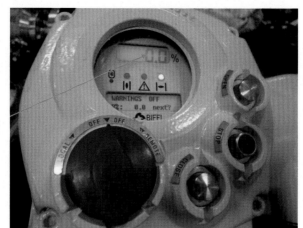

FIG. 2.6

An electrical actuator indicating the fully closed (0% open) position of the electrical actuator and connected valve.

- Electrical actuators are the safest actuators with zero problem for the environment, unlike with oil, and electricity does not have any pressure to damage the human, unlike with oil and air in hydraulic and pneumatic actuators, respectively.
- Electrical actuators typically have a screen on site that shows the percentage of the valve opening. As an example, Fig. 2.6 illustrates 0% opening of the valve on the screen of the electrical actuator connected to the valve.
- No control system and accessories are required for electrical actuators, unlike with both pneumatic and hydraulic actuators.
- Electrical actuators are cheaper, more compact, and lighter in weight when compared to both pneumatic and hydraulic actuators.
- It is possible to monitor the status of the electrical actuators and the connected valves from a long distance, through a control room. Data transfer from the electrical actuators to the control room is typically done by a Process Field Bus (PROFIBUS).
- It is easier to route electricity to the actuators through cables rather than through pipework or tubework, both of which are used for directing and routing air and hydraulic oil to the actuators.
- Electrical actuators have a high torque-to-weight ratio.

PROFIBUS is a digital network that is used to provide communication between the sensors and the control system. The PROFIBUS system for electrical actuators is normally provided by the electrical actuator suppliers. In summary, PROFIBUS provides a means of communication between and control of the electrical actuators. A PROFIBUS design card is placed inside the electrical actuators that are required to be

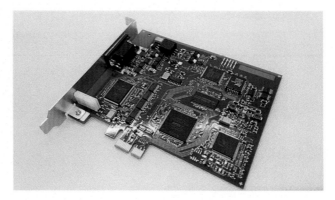

FIG. 2.7

A PROFIBUS card.

controlled, thereby providing the network and wireless protocol or the minimum amount of cable for high-speed data communications from the actuators to the control room. While the PROFIBUS card is inserted into the actuator, the card would be in contact with the electronics inside the actuator, and all the typical commands related to the actuator movement will be transferred into the card. Fig. 2.7 illustrates a PROFIBUS card which is installed inside an electrical actuator. There are wires inside the electrical actuator that transmit data from the electrical actuator to the control room.

As a result, all data as feedback information would be ready in the card. Comprehensive data about the valve and actuator functionality and operational torque values can be transmitted from the electrical actuators to a control room located at a remote distance. The number of actuators that can be connected to a PROFIBUS system varies, but the maximum number of electrical actuators connected to a PROFIBUS system as based on current technology is probably around 120. The status of the actuators connected to a PROFIBUS system is visible in the control room system. In addition, the maximum distance for data transfer between the control room and the actuators is 10 km. Fig. 2.8 illustrates the connection of 8 electrical actuators to the control room through a PROFIBUS data communication system.

Disadvantages of electrical actuators are listed below:

- With the exception of a few specific configurations that are equipped with spring or hydraulic power, electrical actuators cannot provide a failsafe position. Alternatively, fail-as-is or fail-at-the-last-position is achieved by the electrical actuators used for different sectors of oil and gas except for subsea. Subsea electrical actuators can maintain a failsafe position. Additionally, electro-hydraulic actuators can maintain a failsafe mode of operation alternatively.
- Electrical actuators contain more sensitive components compared to pneumatic and hydraulic actuators.

FIG. 2.8

Connection of 8 electrical actuators to a control room through PROFIBUS data communication system.

- Electrical actuators have more complex components internally in comparison with the mechanical parts that are used in pneumatic and hydraulic actuators.
- Electrical actuators are changing due to the advent of new electronic technology. However, the constant development of electrical actuators with new technology can be counted as an advantage.
- Electrical actuators require more documentation and certificates in comparison with both hydraulic and pneumatic actuators, when they are used in hazardous environments such as explosive areas. In addition, electrical components are more hazardous in case of fire as compared to pneumatic and hydraulic actuators.
- Electrical actuators cannot provide stroke speed as fast as pneumatic and hydraulic actuators.

2.2.1.4 Gas

Gas could be used in piston-type actuators for moving the double-acting actuators. Typically, gas is used as a source of power in the actuators in combination with oil. Using gas inside the actuators can cause fugitive emission, so it is not proposed to use this source of power due to damage to the environment. Fugitive emission is defined as an unintentional and undesirable emission, leakage, or discharge of the gases or vapors from pressure-containing equipment or facilities, and from components inside a plant such as valves, piping flanges, pumps, storage tanks, compressors, etc. Fugitive emission is also known as leak or leakage.

2.2.2 Subsea oil and gas

Subsea actuators could be either electrical or hydraulic. Pneumatic actuators are not applicable to subsea valves, since subsea actuators are typically designed for high rated working pressure (RWP) values such as 3000 psi, equal to 207 bar, or can even be standardized to 5000 psi, equal to 345 bar, according to API 17D/ISO 13628-4 standards. Compression of air at such high pressure would be explosive, so

ROV
override

ROV

FIG. 2.9

Subsea valve operation through ROV override by an ROV.

pneumatic actuators are useless for the subsea oil and gas industry. In addition, compressing air at such high-pressure values causes high compression of the air which does not allow for operation of the valve. In general, subsea actuators should be designed to withstand low temperature and high pressure in remote areas. In addition, subsea actuators are normally equipped with a remote operated vehicle (ROV) override. ROV override makes the actuators operable independently from the source of power. In other words, the ROV override can operate the valve in a manner that is independent of the actuator. As an example, it is possible to open and close a hydraulic actuator independently of the hydraulic supply as well as of the spring force. The ROV override is operated manually by an ROV as illustrated in Fig. 2.9. An ROV is an underwater vehicle or a machine that is used to explore the ocean and, more importantly, to perform underwater operations.

The size, dimension, and design of ROV override are typically according to ISO 13628-8 or API 17H standards, which address ROV interfaces on subsea production systems. ROV override is normally installed on the actuator as illustrated in Fig. 2.10, but installation of ROV override on the actuator is not always the case.

2.2.2.1 Hydraulic oil

The concept, design, and type of hydraulic actuators in subsea environments are in general the same as the actuators used in other sectors of oil and gas. Therefore a subsea hydraulic actuator is a cylinder that converts the hydraulic power to mechanical work. The differences between the subsea hydraulic actuators and the hydraulic actuators in other sectors of oil and gas are, as mentioned before, their usage in remote areas and at high pressure and low temperature. Fig. 2.11 illustrates, in yellow color, a subsea hydraulic actuator installed on the top of a subsea valve.

Subsea hydraulic actuators have the main advantages of providing high torque value, fast operation speed, failsafe mode, easy achievement, and robust design suitable for harsh subsea environments with high pressure, low temperature, and remote location. However, the usage of hydraulic actuators has disadvantages such as adding

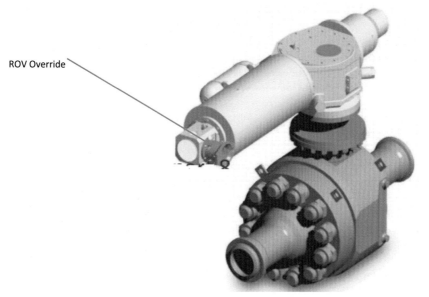

ROV Override

FIG. 2.10

Subsea actuated ball valve with an ROV override.

FIG. 2.11

Subsea hydraulic actuated valve.

extra cost due to subsea hydraulic transportation system, and the possibility of hydraulic oil leakage to the sea and pollution of the environment. Due to the limitations and disadvantages of subsea hydraulic actuators, electrical actuators have been used in the subsea industry for less than two decades.

2.2.2.2 Electricity

The subsea industry is constantly moving toward simplification, cost reduction, and digitalization as the production of oil and gas is moving toward deeper water areas. Using systems that are entirely or all electrical—including the electrical power supply, control, and electrical actuators—instead of hydraulic actuators is one of the main subsea developments in the recent decades. The benefits of entirely electrical systems are health, safety, and environment (HSE) as well as improvement in reliability, when compared to hydraulic systems. Two other advantages of entirely electrical actuators are lower cost and faster response time, when compared to hydraulic actuators. Using entirely electrical subsea systems leads to the elimination of hydraulic control and transition systems, which is a big cost relief in reducing complexity. The cost reduction due to subsea electrical actuators covers both the expenditure and operational costs known as CAPEX and OPEX. In addition, the speed of electricity is faster than the speed of hydraulic systems, which causes a faster response time of electrical actuators when compared to hydraulic actuators. One of the limitations or disadvantages of subsea electrical actuators is the same as that of electrical actuators in other sectors of oil and gas, which is that it is difficult to maintain a failsafe position but it is possible through using either a mechanical or electrochemical spring. It is possible to get feedback from subsea electrical actuator operation, and electrical actuators have the possibility of being equipped with online or condition monitoring. Online valve and actuator monitoring, sometimes called "Valve-Watch," is a state-of-the-art method for improving the safety and reliability of valves and actuators. This well-known method applies some sensors onto the valves and actuators to detect possible failure before the failure becomes more critical. All of the above-mentioned benefits make the electrical actuators and systems well suited to future subsea field development.

2.3 Actuators: Types of design

The motion of the valve stem is divided into linear motion and rotating motion. The design and selection of the actuators should match the type of valve stem motion. Therefore in general, two types of actuators are used: linear and rotary. Linear actuators should be used for linear stem moving valves such as gate and globe. Actuators that can provide rotary motion, such as scotch yoke and rack and pinion, are selected for rotating stem valves, such as ball and butterfly valves.

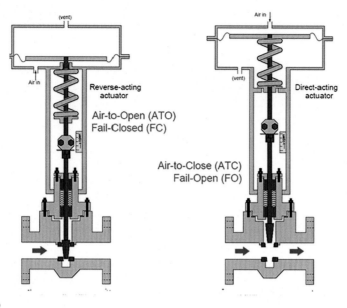

FIG. 2.12

Comparing direct-acting and reverse-acting diaphragm actuators.

Courtesy: Instrumentation tools.

2.3.1 Linear actuators

2.3.1.1 Diaphragm actuators

Diaphragm actuators are pneumatically operated and they use an air supply from the control system or other sources. Diaphragm actuators are normally used for control valves, which are a type of globe valve that is typically used to regulate the fluid in order to adjust some process variables, such as pressure, temperature, or flow rate. Diaphragm actuators are used in all sectors of the oil and gas industry except for subsea. The most common styles for diaphragm actuators are known as "direct-acting" and "reverse-acting." Fig. 2.12 compares direct-acting (right side) and reverse-acting (left side) diaphragm actuators used for the control valves. In direct-acting actuators, the air enters the top area of the diaphragm and pushes the diaphragm down. The air pressure, which is transferred to the diaphragm, overcomes the spring torque located under the diaphragm, so it pushes the valve stem down and closes the valve. Any failure in air supply leads to movement of the stem in an upward direction and the valve opens. Thus direct-acting diaphragm actuators are suitable for air-to-close (ATC) and fail-open (FO) applications. Most of the control valve actuators are reverse-acting, thereby providing a fail-closed (FC) mode of failure. In reverse port diaphragm actuators, the air supply port is located under the diaphragm, so the air supply opens the valve and a stoppage of the air supply leads to closure of the valve.

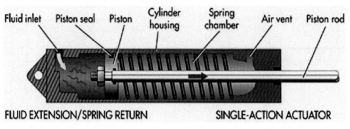

FIG. 2.13

Linear spring return actuator.

The amount of force produced in this type of actuator depends on three main parameters: air pressure, diaphragm diameter, and spring force. Rotary diaphragm pneumatic actuators are the third type of diaphragm actuators and will be discussed later in this chapter.

2.3.1.2 Linear piston actuators

Linear piston actuators can work with either air or hydraulic systems. Fig. 2.13 illustrates a spring return hydraulic actuator with a linear stem movement. Although the actuator in the figure is installed horizontally, the actuator is typically installed in a vertical position on a valve, in the same manner as the actuator (in *red* color) installed on the gate valve in Fig. 2.4. The hydraulic oil enters the chamber on the left and overcomes the spring force, so the piston rod moves to the right side, and consequently the movement of the stem rod is transferred to the valve. Linear piston actuators could be either "single-acting," which is also called "spring return," or "double-acting." Single-acting actuators, as explained earlier, are pressurized by the fluid from one side and return back to the failsafe mode from another side by the spring force. However, double-acting actuators have a springless operation and are pressurized by the fluid from both sides. Double-acting actuators cannot provide a failsafe mode in case of failure of the supply fluid, due to the springless design. The failure mode of double-acting actuators is fail-as-is or staying-in-position.

2.3.1.3 Linear electrical actuators

Electrical actuators can be used on the through conduit gate valves with linear stem movement, as illustrated in Fig. 2.14. The rotary motion in the motor is converted to linear movement. The output thrust in the linear electrical actuator depends on the gear ratio inside the actuator. The connection between the electrical actuator and the valve is a flange connection (see Figs. 2.14 and 2.15), according to ISO 5210 standard, multi-turn valve and actuator attachment. The valves and actuators illustrated in the figure are used for topside and offshore oil and gas industry applications. Subsea electrical actuators have different arrangements for the electrical actuators. A sample design of subsea electrical actuator is reviewed in Chapter 4 of this book.

FIG. 2.14

Linear electrical actuators on the through conduit gate valves with linear stem movement.

2.3.2 Rotary actuators

2.3.2.1 Rotary diaphragm actuators

Like linear diaphragm actuators, this type of actuator works with air and is considered as a pneumatic actuator. Rotary diaphragm actuators are applicable for all sectors of the oil and gas industry except for subsea. Rotary diaphragm actuators have the same mechanism as linear diaphragm actuators on the diaphragm side. Therefore the linear movement is produced initially in the actuator and is converted to rotary motion through a piston-crank mechanism as illustrated in Fig. 2.16. "Piston-crank" refers to converting the linear motion to rotary motion through the connection of a piston with linear movement to a crank with rotary movement through a connecting rod as illustrated in Figs. 2.16 and 2.17.

FIG. 2.15

Flange connection between the valve and the electrical actuator as per ISO 5210.

2.3.2.2 Rotary piston type actuators

There are three types of rotary piston actuators, all of which are discussed in this section: rack and pinion, scotch yoke, and helical spline actuators. All of these types of actuators can be used in all sectors of oil and gas such as subsea, topside, refineries, and petrochemical plants.

Rack and pinion actuators

Rack and pinion is a type of rotary actuator comprising two parts or two types of gears, one of which is a rack with a linear gear movement that moves against the pinion with circular gears. The linear movement of the rack, as illustrated in Fig. 2.18, is transferred through teeth to the gears in the pinion, such that rotary movement in the actuator is transferred to the connected valve through pinion. Rack and pinion actuators work with both hydraulic and air systems, and could be made both as single-acting and double-acting.

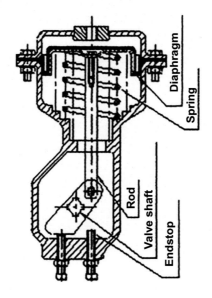

FIG. 2.16

Rotary diaphragm actuator with piston-crank mechanism.

Courtesy: Elsevier.

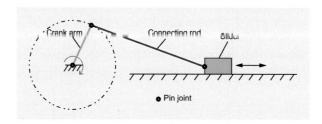

FIG. 2.17

Piston crank mechanism.

FIG. 2.18

Rack and pinion.

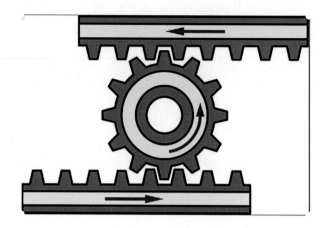

FIG. 2.19

Double rack and a pinion.

Courtesy: Shutterstock.

Rack and pinion actuators may contain a double rack, cylinder, and piston for higher torque output. Fig. 2.19 illustrates double rack and a pinion which is used inside a rack and pinion actuator to provide pressure balance feature for the pinion. The pressure balance feature applied to the pinion. The pressure balance feature refers to using two racks to move pinion from both sides to achieve higher balance in pinion rotation.

Scotch yoke actuators

This is another type of rotary actuator, suitable for quarter turn valves such as ball and butterfly. This type of actuator can work with hydraulic or air systems and could be either the single-acting or double-acting type. Scotch yoke actuators, as illustrated in Fig. 2.20, have a shaft or piston, and the transformation of the linear movement of the shaft into rotary motion is performed through the yoke mechanism.

In this type of actuator, the piston is connected to an output shaft. As the piston moves, due to a fluid force such as air or hydraulic, the linear motion is transferred to the shaft and the yoke mechanism converts the linear movement to rotary movement. There are three main advantages that are associated with scotch yoke actuators: the production of high torque, smooth operation, and few moving parts. Fig. 2.21 illustrates how a hydraulic subsea scotch yoke actuator works. In stage #1, on the top, oil enters the piston and cylinder from the right and compresses the actuator spring and eventually opens the valve. Due to the oil pressure from the right side, the shaft and yoke will be moved to the left side. In stage #2, the oil supply is stopped, so the spring will close the valve. Due to the spring force from the left side, the shaft moves to the left, as does the yoke.

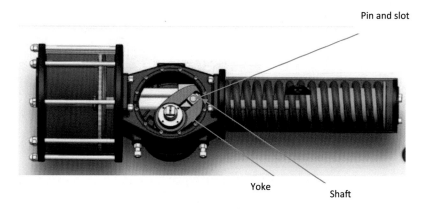

FIG. 2.20

Scotch yoke spring return actuator.

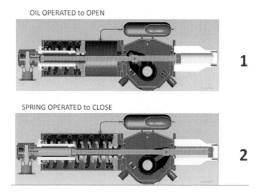

FIG. 2.21

Scotch yoke spring return (single-acting) hydraulic actuator operation.

Helical spline actuators

Helical spline actuators are a type of rotary valve that can be used for quarter turn valves such as the ball valve. One of the reasons why helical spline actuators could be preferred over rack and pinion or scotch yoke actuators is that helical splines stand vertically over the valve, which is connected to a horizontal line. Therefore using helical spline actuators saves a lot of horizontal space that is taken up by both rack and pinion and scotch yoke actuators. In addition, the weight of a helical spline actuator is less than that of two other rotary actuator types. However, helical spline actuators are normally more expensive than both rack and pinion and scotch yoke actuators.

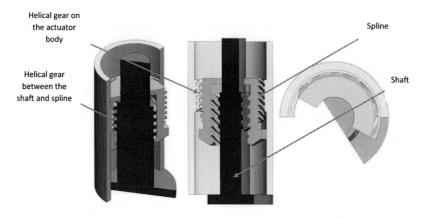

FIG. 2.22

Helical rotary actuator internals.

FIG. 2.23

Helical spline actuator.

In the working mechanism of helical spline actuators, as illustrated in Fig. 2.22, the housing of the actuator has internal gears connected to the external section of the spline. Hydraulic fluid enters from the top and pushes the spline down, while the helical gearing causes the spline and the shaft to rotate at the same time. There are two helical gears: one is between the external part of the spline and the internal section of the actuator body or casing, and the other is between the external section of the shaft and the internal section of the spline. Rotation of the bottom of the shaft will be transferred to the valve. Fig. 2.23 illustrates a helical spline actuator that can be installed vertically on a valve with rotating motion.

FIG. 2.24

Ball valve with an electrical actuator.

2.3.2.3 Rotary electrical actuators

Electrical actuators can provide rotary motion for quarter turn valves such as ball valves. This type of actuator can be installed on different valve sizes, from small sizes of 1" to sizes of 38" or even larger. The output torque can be adjusted through adding more gears or through the gear ratio. The motors in the electrical actuators can work with AC or DC voltage. The housing could be made of different materials such as cast iron, carbon steel, or aluminium. The flange connection between the valve and actuator should be according to ISO 5211 standard, part turn actuator attachments. Fig. 2.24 illustrates a rotary electrical actuator on a ball valve for a topside offshore platform.

2.3.2.4 Rotary electro-hydraulic actuators

Electro-hydraulic actuators are a type of hydraulic actuator that works with the electrical motor. The electrical motor is integrated into the hydraulic actuator as illustrated in Fig. 2.25. Hydraulic actuators could be the piston type, the rack and pinion type, or the scotch yoke type. The output torque of rotary electrical actuators is high and very accurate. Electro-hydraulic actuators are simpler and more reliable than hydraulic actuators, due to elimination of the pump for moving the hydraulic oil and tubing. Electro-hydraulic actuators, unlike electrical actuators, can easily provide a failsafe function.

2.3.2.5 Rotary direct gas actuators

This type of actuator uses high-pressure gas or a nitrogen supply to open or close the valve. Typically, the supplied gas for the actuator is provided by a gas transmission system, such as a gas transport pipeline. Fig. 2.26 illustrates a rotary direct gas scotch

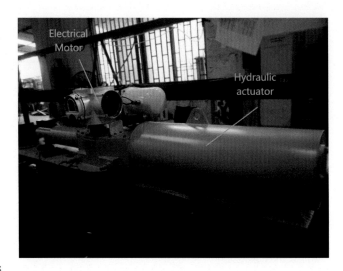

FIG. 2.25

An electro-hydraulic actuator.

FIG. 2.26

Direct gas actuator.

Courtesy: Shutterstock.

yoke actuator that is pressurized by oil from the left side and by gas from the right side. As illustrated in the figure, a direct gas actuator is normally used for double-acting actuators. The use of gas inside a direct gas actuator can be the source of fugitive emission. Thus usage of gas should be avoided to prevent the possibility of fugitive emission.

FIG. 2.27

Gas-over-oil actuator.

Courtesy: Schlumberger.

2.3.2.6 Rotary gas-over-oil actuators

This type of actuator uses gas from the pipeline to pressurize the hydraulic fluid to move a double-acting hydraulic actuator. Using high-pressure gas eliminates the requirement for using pumps and extra tubing to pressurize the hydraulic oil. Fig. 2.27 illustrates a gas-over-oil actuator with two tanks or capsules which are filled with oil and gas pressure over the oil. In the same manner as with the rotary direct gas actuators, using gas inside a gas-over-oil actuator could be a source of fugitive emission. Some of the valves in remote areas, such as those that are located on pipelines, are actuated by natural gas pressure and rotary gas-over-oil actuators.

2.4 Control system and actuator accessories

This section is intended to explain components that are used in the control system or control panel. One should bear in mind that electrical actuators do not have any control panel as a separate unit, so control systems or panels discussed in this section are applicable only to pneumatic and hydraulic actuators. Control accessories in electrical actuators are integrated into the electrical actuators themselves, unlike with pneumatic and hydraulic actuators. In fact, electrical actuators do not require any space for a control system as a separate unit, which can be counted as an advantage. Another important point is that the control components discussed in this section

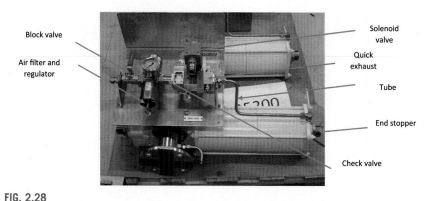

FIG. 2.28

Control panel on a pneumatic actuator.

are applicable to the actuators in other sections of oil and gas, except subsea. Fig. 2.28 illustrates a control panel on a pneumatic scotch yoke actuator for a butterfly valve.

The same arrangement for a control system is applicable to hydraulic actuators, but, as mentioned earlier, control panels for hydraulic actuators are larger.

2.4.1 Block valve

Block valves, which could be ball or needle valves, are used to make the source of energy (air or hydraulic) move in the control panel in order to operate the actuator. The block valve illustrated in Fig. 2.28 is a ball valve, which is in the closed position. Closure of the ball valve isolates the source of energy such that the actuator returns back to either a fail-open or closed position upon losing the air supply. But the question is: how is it possible to figure out the position of the ball valve? The position of the lever of the ball valve, in blue color, shows the direction of the hole inside the ball. Since the lever of the ball valve is perpendicular to the flow of air in the control panel, the hole of the ball in the ball valve should be perpendicular to the flow of air in the control panel, which means that the ball valve is in the closed position. The block valve is normally not the only valve on the control panel. The check, solenoid, and quick exhaust valves (applicable only to pneumatic actuators) are also located on a control panel, and will be discussed later in this chapter.

2.4.2 Filter and regulator

This component is applicable to both pneumatic and hydraulic actuators, for removing dust and dirt, such as particles and rust from the air or hydraulic oil. The cleanliness of both air and hydraulic oil is important, since having particles in the actuator can disturb the functionality of the actuators. The cleanliness classification for the hydraulic fluid could be defined according to different standards such as SAE

4059. The filter is integrated with a regulator to adjust the pressure of the supply fluid to the actuator. As shown in Fig. 2.28, it is possible to see and measure the pressure of the supply fluid to the actuator.

2.4.3 Check valve

The check valve is a nonreturn valve that works in one direction. The check valve opens when the supply fluid passes through it from the left side, but the valve does not allow the fluid to return back to the upstream side of the valve after having passed through the valve.

2.4.4 Solenoid valve

This valve is an electromechanically operated valve that gets opened and closed through the starting or stopping of electricity. Solenoid valves are used for both pneumatic and hydraulic actuators. The number of solenoid valves on the control panel, as illustrated in Fig. 2.28, is just one; however, the number of solenoid valves on the control panel could be two or three, if higher safety functions, such as an emergency shut down feature, are required. Fig. 2.29 illustrates the functioning of a solenoid valve. The figure on the right shows a condition in which electricity is supplied to the solenoid valve, so the coil is energized and a magnetic field is created which encompasses the plunger, in blue color. Movement of the plunger to the top, due to the coil magnetic force, pulls the plunger up, so the valve opens. Loss of electricity eliminates the magnetic field, so the plunger moves down and the valve closes, as shown on the left side of the figure.

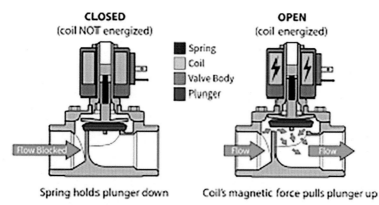

FIG. 2.29

Solenoid valve functioning.

2.4.5 Quick exhaust valve

Quick exhaust valves are used on the control panel in order to provide rapid piston speed that is achieved on the return stroke of single- and double-acting pneumatic actuators. Quick exhaust valves are not used on the control panels of hydraulic actuators, since there is the presence of hydraulic oil, which is not environmentally friendly, unlike air.

2.4.6 Limit switch

A limit switch is not located on the control panel, so it is not illustrated in Fig. 2.28. Limit switch boxes, as shown in Fig. 2.30, are sensors installed on pneumatic and hydraulic actuators to control the motion of the actuators. One of the main functions of a limit switch is to make sure the device does not extend its mechanical movement range. As an example, a limit switch should not allow the movement of an actuator to occur in such a manner as to rotate the ball of a ball valve connected to the actuator more than 90 degrees.

2.4.7 Tube

Typically, a small tube of maximum 1″ size is used to connect the control panel to the actuator. The components located on the control panel could be connected through tubes or without a tube connection. The components on the control panel are connected like blocks without any tube connections. Tubing material could be austenitic stainless-steel grade 316 or 6Molebdeneym (6MO) alloy. 6MO is a superaustenitic alloy with much higher corrosion resistance and mechanical strength compared to stainless steel 316. Tubes may be supported in certain lengths by some clamps. Tubes located under the clamps are subject to stress cracking corrosion in offshore environments, which is why the tubing material could be upgraded to 6MO from stainless steel 316 in some cases where the tubing is supported through clamps and in offshore oil and gas.

Limit switch box

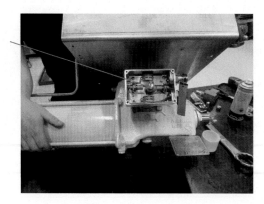

FIG. 2.30

Limit switch on an actuator.

Position indicator

FIG. 2.31

Actuator position indicator shows open position.

2.4.8 End stopper

The end stopper is not located on a control panel. The end stopper is located at the end of the actuator cylinder that is used to completely adjust the valve closure member in the correct position. As an example, the ball of a ball valve is moved 89 degrees to the closed or open position by the actuator. The one degree percent offset of the ball can be fixed through rotating the end stopper on the actuator.

2.4.9 Position indicator

The position indicator is located on the actuator and not on the control panel. It shows the position of the actuator. As an example, the position indicator illustrated in Fig. 2.31 indicates that the actuator and the connected valve are in the open position. Proximity switches may be installed on some of the actuated valves to report the position of the actuators to the control room.

Questions and answers

1. Which types of actuators are not environmentally friendly?
    A. Hydraulic actuators
    B. Electro-hydraulic actuators
    C. Pneumatic and electrical actuators
    D. A & B are correct

Answer: Hydraulic actuators are not environmentally friendly. The spilled hydraulic oil can be absorbed by the ground and it can contaminate the water as well as food. Electro-hydraulic actuators are also not considered environmentally friendly; in fact, electro-hydraulic actuators are a type of hydraulic actuator in which the hydraulic oil as a source of energy is moving with electrical motors rather than pumps. Pneumatic and electrical actuators are both safe for the environment. Thus, option D is the correct answer.

2. Which type of actuator does not have a control panel?
  A. Electrical actuators
  B. Pneumatic actuators
  C. Hydraulic actuators
  D. Hydro-electrical actuators

Answer: Electrical actuators do not have a control panel. Thus option A is correct. Pneumatic and hydraulic actuators have control panels. Hydro-electrical actuators are a type of hydraulic actuators, so they have control panels.

3. Which sentences are correct about electro-hydraulic actuators?
  A. Electro-hydraulic actuators are a type of electrical actuator that work with electrical power.
  B. The reliability of electro-hydraulic actuators is less than hydraulic actuators due to a simpler system architecture.
  C. Using electro-hydraulic actuators eliminates the need for a separate pump and extra tubing.
  D. Electro-hydraulic actuators are a type of hydraulic actuator that work with hydraulic fluid.

Answer: Option A is not correct, since electro-hydraulic actuators are not a type of electrical actuator. Option B is not correct, since the reliability of electro-hydraulic actuators is higher than that of the hydraulic actuators. Option C is correct, because using electro-hydraulic actuators eliminates the need for a separate pump and extra tubing. Option D is also correct, since the electro-hydraulic actuators are a type of hydraulic actuator that work with hydraulic fluid. Fig. 2.32 illustrates a series of

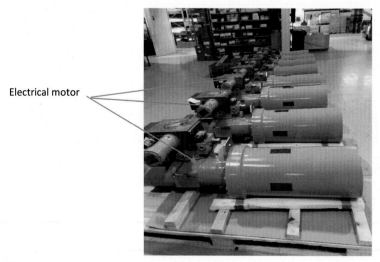

Electrical motor

FIG. 2.32

Electro-hydraulic actuators highlighting the electrical motor part.

electro-hydraulic actuators in a valve factory. Each hydraulic cylinder or actuator has an electrical motor attached to it. Thus, there is no need to have a separate pump or extra tubing.

4. Which type of actuator is proposed for "ease of operation"?
    A. Double-acting hydraulic actuators, since the force produced by the hydraulic actuators is higher than that of air or electricity.
    B. Electrical actuators, since this type of actuator is equipped with manual override.
    C. Double-acting actuators, since they do not have any spring.
    D. Electrical actuators, since they have a motor that facilitates the valve operation, compared to a manual option with a handwheel or lever.

Answer: Option A is not correct, since double-acting hydraulic actuators are used for producing high torque as well as fast speed operation. Option B is not correct, since the option of manual override on an electrical actuator does not provide ease of valve operation. Option C is not correct, since double-acting actuators are not selected due to ease of operation. Option D is correct, because electrical actuators are used for ease of operation, and the presence of a motor in electrical actuators facilitates fast operation of the valve when compared to manual valve operation choices, such as a lever or handwheel.

5. Which components are not located on a pneumatic control panel?
    A. Limit switch, tubing, quick exhaust
    B. Limit switch, position indicator
    C. Pressure gauge, filter and air regulator
    D. Solenoid and block valve

Answer: Tubing, quick exhaust, pressure gauge, filter and air regulator, and solenoid and block valves are all located on a pneumatic actuator control panel. However, a limit switch and a position indicator are located outside the control panel, so option B is the correct answer.

6. Two types of actuators are illustrated in Fig. 2.33. Which sentence is correct about these two types of actuators?
    A. These actuators are both rotary type and single-acting type.
    B. The one on the top is a scotch yoke actuator and the one on the bottom is a rack and pinion type; but both actuators have linear movement.
    C. Both actuators are double-acting with linear movement.
    D. The one on the top is a scotch yoke actuator and the one on the bottom is a rack and pinion type; both have rotary type actuators and they are both double-acting.

Answer: The actuator on the top is a scotch yoke type and the one on the bottom is a rack and pinion type. Both actuators do not have any spring, and they should be opened and closed by the power of the fluid supply, which could be air or hydraulic oil, so they are double-acting actuators. Both types of actuators are rotary type, since

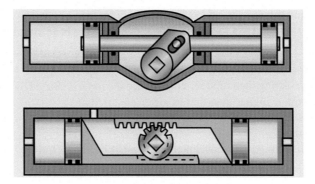

FIG. 2.33

Two types of actuators.

Courtesy: Valve magazine.

the linear motion is converted to rotary motion by the yoke and connection pin on the scotch yoke actuator and by the pinion in the rack and pinion type actuator. Thus option D is correct.

7. Which types of actuators could be the source of fugitive emission?
    A. Hydraulic rack and pinion actuators
    B. Gas-over-oil as well as direct gas actuators
    C. Linear and rotary electrical actuators
    D. Pneumatic rack and pinion actuators

Answer: Gas is the only source of fugitive emission in actuators. Hydraulic oil spill in the environment is damaging the environment but is not considered as a source of fugitive emission. Air and electricity are 100% clean sources of energy for the environment. Therefore hydraulic rack and pinion actuators, linear and rotary electrical actuators, and pneumatic rack and pinion actuators are not a source of fugitive emission. Option B is the correct answer, since both gas-over-oil and direct gas actuators utilize gas, so they are sources of fugitive emission.

8. The pressure of the oil supplied to a failsafe closed hydraulic actuator is very low, below the minimum design for oil pressure, which is 160 bar. Which sentence is correct regarding this condition?
    A. The valve cannot be closed fully.
    B. The solenoid valve could be the reason for the low pressure of the oil.
    C. The valve is not getting opened fully.
    D. The exhaust valve could be less exposed to wear due to low oil pressure.

Answer: The valve and actuator are failsafe closed, which means that they are getting closed by the spring force, so low pressure of the oil should not cause any problem for closing the valve and the actuator. The solenoid valve does not have anything to do with pressure of the oil. Therefore both options A and B are not correct. However, low pressure of oil in the failsafe closed valve could prevent the valve and the connected actuator from fully opening. Thus option C is correct. The exhaust valve is not used in hydraulic actuators, so option D is also not correct.

9. Which sentence is not correct about actuators?
A. Pneumatic and electrical actuators are more environmentally friendly than hydraulic actuators.
B. Both rack and pinion and scotch yoke could be either pneumatic or hydraulic.
C. Failsafe closed actuators are closed with a spring force.
D. Changing an actuator for a specific valve from the pneumatic to the hydraulic type leads to a larger actuator.

Answer: Option A is correct, because hydraulic actuators are less environmentally friendly than both pneumatic and electrical actuators. Option B is correct, since both rack and pinion as well as scotch yoke could be either pneumatic or hydraulic. Option C is correct, because a failsafe closed actuator is closed with a spring force. Option D is not correct, since hydraulic actuators are more compact than pneumatic actuators for a specific valve.

10. Which sentence is not correct about the control systems of pneumatic actuators?
A. Different types of valves, such as isolation valve, check valve, solenoid valve, and exhaust valve, are located on a pneumatic actuator control panel.
B. The components on the control panel could be connected with tube or without tube.
C. Air filter and regulator can be integrated into one component on the control panel.
D. The connection between the control panel and the actuator is tubeless.

Answer: All the options are correct except option D. Typically, a small-size tube such as ½″ is connected between the actuator and the control panel. Fig. 2.34 illustrates the tube connection between the control panel and the pneumatic actuator.

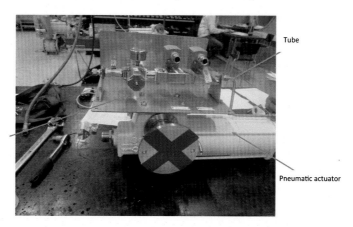

FIG. 2.34

Tube connection between the control panel and pneumatic actuator.

Further reading

American Society of Mechanical Engineers (ASME) B16.34, Valves—Flanged, Threaded, and Welding End, American Society of Mechanical Engineers, NY, USA, 2017.

American Petroleum Institute (API) 17D, Design and Operation of Subsea Production Systems, Subsea Wellhead and Tree Equipment, second ed., American Petroleum Institute, Washington, DC, USA, 2011.

American Petroleum Institute (API) 17H, Remotely Operated Tools and Interfaces on Subsea Production Systems, third ed., American Petroleum Institute, Washington, DC, USA, 2019.

Fisher Controls International (Ed.), Control Valve Handbook, third ed., Fisher Controls International, Marshalltown, IA, USA, 2001.

International Organization of Standardization (ISO) 13628-4, Petroleum and Natural Gas Industries—Design and Operation of Subsea Production Systems—Part 4: Subsea Wellhead and Tree Equipment, second ed., International Organization of Standardization, Geneva, Switzerland, 2010.

International Organization of Standardization (ISO) 13628-8, Petroleum and Natural Gas Industries—Design and Operation of Subsea Production Systems—Part 8: Remotely Operated Vehicle (ROV) on Subsea Production Systems, first ed., International Organization of Standardization, Geneva, Switzerland, 2006.

International Organization of Standardization (ISO) 5210, Industrial Valves—Multi Turn Valve Actuator Attachments, second ed., International Organization of Standardization, Geneva, Switzerland, 2017.

International Organization of Standardization (ISO) 5211, Industrial Valves—Part Turn Actuator Attachments, second ed., International Organization of Standardization, Geneva, Switzerland, 2017.

T. Mahl, The new generation of electrical actuators, Valve World Mag. 14 (4) (2009) 49–51.

J. Onditi, E. Carey, Valve Actuation: The when, how and why of Actuator Selection. A Guide to Actuators for Upstream and Midstream Oil and Gas Applications, Flowserve, 2018.

K. Sotoodeh, Actuator selection and sizing for valves, Springer Nature Appl. Sci. (2019), https://doi.org/10.1007/s42452-019-1248-z.

K. Sotoodeh, The actuators that drive subsea operations, Valve Mag. (2020). [online]. Available at: http://www.valvemagazine.com/magazine/sections/features/10872-the-actuators-that-drive-subsea-operations.html. (Accessed 26 November 2020).

K. Sotoodeh, All-electrical subsea control systems and the effect on subsea manifold valves, J. Marine Sci. Appl. (2020), https://doi.org/10.1007/s11804-020-00155-1.

Actuator selection for reducing emission

3.1 Actuators and fugitive emission

Fugitive emission is defined as the unintentional and undesirable emission, leakage or discharge of gases or vapors from pressure-containing equipment or facilities and components inside plants such as valves, piping flanges, pumps, storage tanks, valves, compressors, etc. Fugitive emission is also known as leak or leakage. The term 'fugitive' is used because these emissions are not taken into account and calculated during the design of the equipment and components. Fig. 3.1 illustrates fugitive emission issuing from a flare tip inside a refinery.

There are many negative impacts associated with fugitive emission, such as environmental pollution. The *greenhouse effect* is a process in which fugitive emission gases in the earth's atmosphere trap the sun's heat. This process causes the earth to be much warmer than it should be. In fact, *global warming* (see Fig. 3.2) is one of the negative consequences of the greenhouse effect. Particular gases, such as carbon dioxide (CO_2), methane (CH_4), etc., that contribute to the greenhouse effect and thus to global warming are called "*greenhouse gases.*"

The negative effects of fugitive emission compounds, such as volatile organic compounds (VOCs) and hazardous air pollutants (HAPs), are not limited to global warming and the greenhouse effect. Other environmental effects of VOC and HAP emission include photochemical ozone creation potential (POCP), ozone depletion potential (ODP), toxicity, carcinogenicity and local nuisance from odor. Ozone layer depletion increases the amount of ultraviolet (UV) sun radiation that reaches the earth's surface, which can cause skin cancer. Ground-level ozone creation occurs when emissions from cars, refineries, plants and other industrial units react chemically in the presence of sunlight. Thus ground-level ozone can reach unhealthy levels, especially on warm days. Inhaling ozone can damage the lungs and cause chest pain, coughing, shortness of breath and throat irritation.

Humans have been polluting the environment for many years, especially since the advent of the industrial revolution in the 19th century. A major increase in fugitive emission occurred in the 1950s when the world began to require more heat and electricity. It is noticeable that the amount of fugitive emission in 2020 is almost five times more than it was 70 years ago. Fugitive emission has long been known as the key concern for end users and operators of oil and gas, chemical and petrochemical plants, as well as regulators across the globe. A variety of reasons and factors, such as regulations from governments, health, safety and environment (HSE)

Prevention of Actuator Emissions in the Oil and Gas Industry. https://doi.org/10.1016/B978-0-323-91928-9.00001-3

FIG. 3.1

Fugitive emission in a refinery issuing from a flare tip.

Courtesy: Shutterstock.

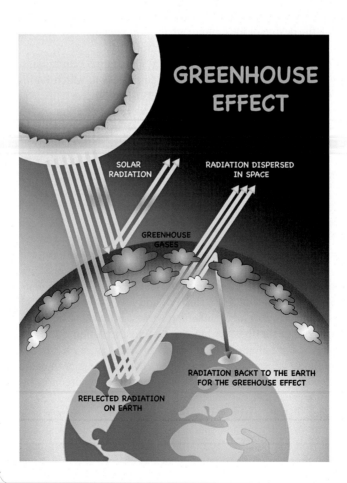

FIG. 3.2

Greenhouse effect and global warming.

Courtesy: Shutterstock.

programs and increasing pressure from the public have forced end users in the oil and gas industry to be more conscious of and conscientious about preventing fugitive emission.

The term 'fugitive emission' is not applicable to actuators in most cases, except when gas is used inside them. Usage of high-pressure gas in actuators is limited to rare cases where direct gas actuators or gas over oil actuators are used. As an example, most of the actuators which are used in the offshore industry are powered by air, hydraulic oil and electricity. Air emission from actuators is costly, but it is not considered as a fugitive emission. Hydraulic oil emission to the environment pollutes the environment and is against HSE policies, but it is not considered as a fugitive emission either. As illustrated in Fig. 3.3, actuators are not included in the list of components and facilities responsible for or involved in fugitive emission. Thus unlike industrial valves, no fugitive emission test has been developed for actuators. Therefore, it can be concluded that the actuators are not considered as a source of fugitive emission in almost all cases except for rare cases where gas is used inside them. Thus the recommendation of author is to avoid using actuators which are working with gas to prevent fugitive emission. Therefore there is no need for the actuators to be tested for fugitive emission detection unlike industrial valves.

It is important to bear in mind that two main strategies are associated with environmentally friendly actuators; the first approach is associated with actuator selection and the second is related to actuator design. Regarding the first strategy, it is important to avoid selecting hydraulic- and gas-containing actuators as much as practically possible. Alternatively, pneumatic and electrical actuators are harmless to the environment. In fact, both air and electricity, which are used as sources of power in pneumatic and electrical actuators, respectively, are environmentally friendly. The focus of this chapter is on selecting environmentally friendly actuators based on the first strategy. It is easy to avoid using actuators that work with high pressure gas in offshore and nonremote areas, but avoiding the usage of hydraulic actuators in the nonsubsea sector of the oil and gas industry is not possible in some cases, such

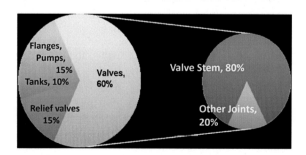

FIG. 3.3

Fugitive emission percentages from different facilities and components in refineries in the oil and gas industry.

Courtesy: Valve World magazine.

as when a large, high-pressure class valve must be actuated. Alternatively, shifting from hydraulic to electrical actuators in the subsea oil and gas industry is possible and is discussed in Chapter 4. The second approach or strategy is to provide more environmentally friendly solutions in terms of the design, selection and testing of hydraulic actuators and relevant hydraulic systems and connections to prevent the leakage of hydraulic fluid into the environment.

3.2 Actuators and hydraulic oil spills

The negative impacts of gaseous fugitive emission on the environment were reviewed in the previous section. It is rare to use an actuator in the oil and gas industry that works with high-pressure gas, so it may be concluded that actuators are not a source of fugitive emission inside oil and gas plants as per the statistics illustrated in Fig. 3.3. Thus the main source of energy for the actuators that can cause environment pollution is hydraulic fluid. This section explains the negative environmental impacts of hydraulic oil spills in more detail. The spillage of hydraulic oil to the environment from actuators occurs through leakage. If hydraulic oil spills onto soil, some fluid may stay on the surface of the soil while some may seep into the ground water. Both surface and underground oil spills have the potential to contaminate soil, sediment, water and air. Contamination of the soil contaminates food production. Food and water contamination can cause serious health problems for a person who eats the polluted food or drinks the polluted water. Spilled hydraulic oil has negative effects on wildlife; when animals ingest hydraulic oil, or when it comes into contact with their skin, it can damage their health. Contact with hydraulic oil can put human health at risk too. As many of approximately 26,000 marine mammals can be harmed by oil spill. At least four species can be killed by oil spill including dolphins and sperm whales. Fig. 3.4 illustrates the leakage of hydraulic oil from hydraulic tube.

3.3 Actuator selection

3.3.1 Nonsubsea oil and gas industry

Many parameters affect actuator type selection, such as the required torque amount, the type, size and pressure class of the connected valve, the speed of operation, the actuator's expected failure mode, such as "fail safe" or "fail-as-is," the availability of the source of power, the cost, whether the installation area is hazardous, etc. Fig. 3.5 summarizes and illustrates the parameters affecting actuator selection. These are also listed below:

1. *Required torque amount:* Torque is defined as any force that tends to create mechanical rotation. In a simple definition, torque is the quantity of force required to rotate a mechanical component around a specific point. Each type of valve requires a specific amount of torque to move between open and closed

FIG. 3.4

Leakage of hydraulic oil from hydraulic tube.

Courtesy: Shutterstock.

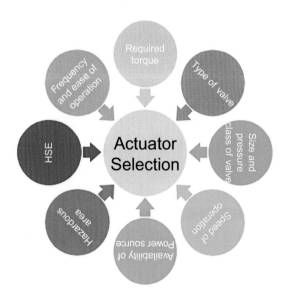

FIG. 3.5

Parameters affecting actuator selection.

positions; this value is provided by the valve manufacturer. The five torque values defined for the operation of a valve are explained as follows:

1. Break to open (BTO): This torque is measured when the valve is closed and the ball opens against just one seat under pressure. This torque is also called breakaway torque.

2. Running torque (RT): The torque of the valve when the ball opens at approximately 35–45 degrees.
3. End to open (ETO): The torque of the valve when the ball is almost 80 degree open to fully open or 90 degree open.
4. Break to close (BTC): When the valve is in fully open position, the torque required to break the open position of the valve to close the valve.
5. End to close (ETC): The torque required to fully close the valve when the valve is about to close.

Fig. 3.6 illustrates how the position of the ball in a ball valve corresponds to each type of torque.

The same torque definitions are used for actuators. The torque produced by an actuator should always be higher than the torque required by the valve. The ratio of actuator torque to valve torque is defined as the safety factor. The safety factor could be minimum 1.5 or 2 which is defined in the valve and actuator specification. Typically, hydraulic actuators produce the highest torque, due to the high pressure of the oil, compared to pneumatic actuators and even electrical actuators of the same size. Table 3.1 illustrates a torque table for an 8″ class 150 butterfly valve with a pneumatic actuator.

According to the table, the actuator should be able to fully open the valve from fully closed position or vice versa in 16 s. The required safety factor of the actuator is 1.5 as a minimum, but in practice and reality the actuator could provide a safety

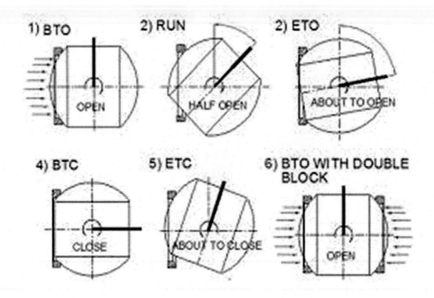

FIG. 3.6

Position of the ball in a ball valve corresponding to each type of torque.

Table 3.1 Torque table for an 8″ class 150 butterfly valve with a pneumatic actuator.

Item	Valve tag	Valve break to open	Valve running	Valve fully open	Valve break to close	Valve running	Valve end to close
1	21XV0005	320.0	176.0	80.0	80.0	176.0	308.0
Displacement	Actuator model	Actuator break to open	Actuator running	Actuator fully open	Actuator break to close	Actuator running	Actuator end to close
7.2	Model X	1463.1	698.6	910.2	1618.1	818.8	1145.2
Valve/size/class	Open/close time/safety factor	Real safety factor	Real safety factor	Real safety factor	Real safety factor	Real safety factor	Real safety factor
8″#150 butterfly valve	16.0/16.0/1.5	4.57	3.97	11.38	20.23	4.65	3.72

factor of a minimum of 3.72 or a maximum of 20.23 (refer to the torque table). The high safety factor values, which are much above 1.5, indicate that the actuator is probably overdesigned and that a smaller actuator could be selected alternatively. The valves' torque values as per the table are 320 Nm (newton-meter), 176 Nm, 80 Nm, 80 Nm, 176 Nm and 308 Nm. The actuators' torque values corresponding to each valve torque value are 1463.1 Nm, 698.6 Nm, 910.2 Nm, 1618.1 Nm, 818.8 Nm and 1145.2 Nm. The reason why the valve and actuator torque values are emphasized is that the safety factors are calculated based on the corresponding valve and actuator torque values. The torque required for operating the valve does not necessarily affect the type of actuator; the same type of actuator in a smaller or larger size can adjust the torque output.

2. *Type of valve:* Different valves have different torque values. As an example, more torque is required to operate ball valves compared to butterfly valves; Table 3.2 compares the torque values for the operation of a ball and butterfly valve, both 10″ in size and in pressure class of 150. Butterfly valves in the Norwegian offshore industry are mainly used for low pressure class applications. Lower pressure class valves require less torque for operation. Thus pneumatic actuators can produce sufficient torque for all butterfly valves with a fail safe mode of operation, from small sizes such as 4″ up to 30″. However, a higher torque value is required for ball valves and hydraulic actuators may be required for some ball valves, such as 20″ and larger with a fail safe mode of operation.
3. *Size and pressure class of the valve:* Generally, valves of larger size and in higher pressure classes require higher torque for operation. As an example, a 3″ class 1500 ball valve with a fail safe mode of operation can be actuated with a pneumatic actuator. However, a pneumatic actuator cannot produce enough torque for the operation of a 42″ ball valve in a pressure class of 1500 with a fail safe mode of operation, so in the second case, a hydraulic actuator is required. Fig. 3.7 illustrates a ball valve in 4″ × 3″ class 300 with a fail safe close mode of operation actuated with a pneumatic actuator and a control panel.
4. *Speed of operation:* Hydraulic actuators provide the fastest speed of operation compared to pneumatic and electrical actuators. Pneumatic actuators have the second ranking of operation speed after hydraulic actuators. Electrical actuators

Table 3.2 Torque comparison between ball and butterfly valves in newton-meter for 10″ size and pressure class 150.

Valve	BTO (Nm)	ETO (Nm)	Running (Nm)	BTC (Nm)	ETC (Nm)
10″ Butterfly valve class 150	585	146.3	293	146.3	527
10″ Ball valve class 150	1242	1039	837	1217	1039

Control panel

Pneumatic
actuator

A 4″x3″ ball valve
class 300

FIG. 3.7

A 4″ x 3″ ball valve in class 300 with a pneumatic actuator and a control panel.

have the slowest speed among all these three types of actuators. Actuated valves that are connected to an emergency shut down system (ESD) should close the connected valve very quickly when a dangerous event is detected. Electrical actuators are not suitable for the actuation of emergency shut down valves, since electrical actuators cannot provide the fast speed of operation required for ESD valves. Alternatively, pneumatic and hydraulic actuators are typically selected for ESD valves. The speed of operation for pneumatic and hydraulic actuators could be as fast as 1″ per second or even faster. As an example, a pneumatic actuator could operate a 10″ valve in 10 s. However, electrical actuators could not operate a valve as quickly as 1″ per second. It would take an electrical actuator installed on a 30″ butterfly valve at least 75 s to open or close the valve. Fig. 3.8 illustrates a 6″ ESD ball valve in pressure class 300 with a pneumatic actuator.

5. *Availability of power source:* In one recent topside offshore project, all the facilities were located on a floating, production, storage and offloading (FPSO) ship; all the processing and production of oil, gas and water took place in the facilities located on the ship. No source of hydraulic power was available on the ship, so all the actuators in the project were selected as either electrical or pneumatic.

6. *Hazardous area:* Electrical actuators require more documentations and certificates compared to both hydraulic and pneumatic actuators when they are used in hazardous environments such as explosive areas. In addition, electrical components are more hazardous in case of fire compared to pneumatic and hydraulic actuators. A hazardous area is defined as any location where there is a

Pneumatic actuator

Emergency shut down 6"
CL300 ball valve

FIG. 3.8

6″ CL300 ESD ball valve with a pneumatic actuator.

risk of explosion. Typically, three zones are considered for vapors, mists and gases. Zone 0, which has the highest risk of fire and exposition, is defined as a location where flammable gases are present continuously for a long period of time. Zone 1, which has a medium risk of fire and explosion, is defined as a location where flammable gases or vapors are generated during regular operational conditions. Zone 2 has the lowest risk of fire and explosion; here, flammable gases are unlikely to be present during regular operation conditions, or if they do occur, they exist for a short time.

7. *Frequency and ease of operation:* Electrical actuators are used in many cases for ease of operation. It is common to use electrical actuators for specific, large sizes of valves in high pressure classes that are operated frequently, rather than to operate these valves manually. As an example, the operation of a 20″ ball valve in pressure class 300 is difficult for a manual operator. Thus an electrical actuator is installed on the valve for ease of operation. Frequency of operation is an important parameter for selecting an electrical actuator for ease of operation. As an example, if the 20″ ball valve mentioned above required operation only once every three years, then the valve could be supplied manually without any actuator.

8. *HSE:* The most environmentally friendly choices of actuators are electrical and pneumatic. Hydraulic actuators are not environmentally friendly. The next section discusses in detail how to select an actuator in the oil and gas industry to reduce fugitive emission and prevent damage to the environment. When it comes to health and safety, probably none of the actuators are completely safe, since working with high pressure hydraulic oil and air as well as electricity is dangerous.

3.3.2 The subsea oil and gas industry

Actuator selection for the subsea oil and gas industry is more straightforward. Traditionally, hydraulic actuators have been used for the subsea valves located on subsea manifolds and Christmas trees. However, newer technology has shifted the subsea industry toward all electrical subsea control and actuators. Table 3.3 indicates a list of the actuated valves in a subsea project. All the actuators are hydraulic with Remote Operated Vehicle (ROV) override operation possibility. ROV override installation on the subsea actuators provides possibility of actuators operation independent of the hydraulic fluid. The hydraulic actuators could be single acting (spring return), or double acting for larger valves in order to provide higher torque. Double acting hydraulic actuators with fail-as-is (FAI) failure mode of operation are selected for 22″ and some of the 16″ ball valves, refer to the table. In addition, it is noticeable that a small size gate valve as small as ¾″ gate valve is selected as an actuated valve due to frequency of operation in very deep area of the water as deep as more than 2 km. The ¾″ gate valve is used to inject the chemicals into the process line to prevent possible operational problems in the process fluid.

3.4 Actuator selection to reduce emission

Many efforts are being made worldwide to reduce emissions from industrial activities, as illustrated in Fig. 3.9. Although actuators are not listed as a source of emission in Fig. 3.3, it is important to know that many of the valves connected to pipelines in remote areas are actuated by natural gas. However, the usage of actuators that work with natural gas is not common in the offshore industry, including both topside

Table 3.3 List of actuated valves in a subsea project.

Valve type	Bore size (mm)	NPS (″)	Operator	Fail mode
Ball valve	476.4	22	Hydraulic with ROV override	FAI
Ball valve	344.4	16	Hydraulic with ROV override	FAI
Ball valve	344.4	16	Hydraulic with ROV override	FAI
Ball valve	344.4	16	Hydraulic with ROV override/spring return	FSC
Ball valve	230.2	10	Hydraulic with ROV override/spring return	FSC
Gate valve	179.4	8	Hydraulic with ROV override/spring return	FSC
Gate valve	46	2	Hydraulic with ROV override/spring return	FSC
Gate valve	19	3/4	Hydraulic with ROV override/spring return	FSC

FIG. 3.9

Undesirable fugitive emissions from industrial activities must be reduced.

and subsea. Considering that fugitive emission can be caused by actuators working with high-pressure gas, such as gas over oil double acting and rotary actuators, the usage of these types of actuators should be eliminated to provide a clean environment. Hydraulic actuators are also a source of hydraulic oil spillage and with it, the negative consequences discussed earlier in this chapter. Unlike hydraulic and gas, air and electricity are clean sources of energy with zero emission to the environment. Although air is a clean source of energy, unlike gas and hydraulic oil, pneumatic actuators are not as economical or efficient as electrical actuators. Environment Protection Agencies (EPAs) propose using electrical actuators for 100% emission reduction wherever applicable. In addition, electrical actuator technology is improving significantly, so modern electrical actuators can provide excellent performance and are ideal for a variety of process applications.

The question is, is it possible to select only electrical actuators in plants? When it comes to subsea actuators, it is possible to shift totally from hydraulic actuators to electrical actuators. But it is not possible to select electrical actuators for all the valves in refineries and chemical plants, or in the topside, offshore sector of the oil and gas industry. The reason why electrical actuators cannot be selected for all the valves in the nonsubsea oil and gas industry refers back to the two main limitations of electrical actuators; the first one is the difficulty of maintaining a failsafe position, and the second is the nonsuitability of electrical actuators for fast operation, which is required in an emergency shutdown (ESD) or high integrity pressure protection system (HIPPS). If the valves require a failsafe mode or a fast speed of operation, a pneumatic actuator should be the first choice unless the valve requires high torque, which cannot be produced by pneumatic actuators, or a higher speed of operation, especially in failure mode, than can be produced by a pneumatic actuator. In the two above cases, a hydraulic actuator is proposed for valve operation. The flow chart illustrated in Fig. 3.10 shows criteria for environmentally friendly actuator selection for all sectors of the oil and gas industry except subsea. Three important points should be kept in mind for the selection of environmentally friendly actuators, according to the flow chart provided below. The first is that no double acting

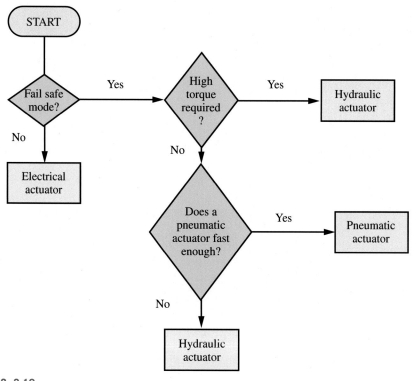

FIG. 3.10

Environmentally friendly actuator selection for all sectors of the oil and gas industry except subsea.

pneumatic or hydraulic actuators should be used in plants. Double acting actuators, as explained in Chapter 2, are fail-as-is, meaning they stay in their last position in the event of failure. Double acting actuators may be used since they produce higher torque and, in general, are more compact than single acting or spring return actuators. The second is that using actuators with high-pressure gas is not proposed at all. The third is that electrical actuators are mainly used for easy operation of valves without any safety function such as ESD, HIPPS or process shutdown.

Environmentally friendly actuator selection for the subsea sector has a more straightforward process. In fact, all hydraulic actuators should be replaced with electrical actuators in the subsea sector in order to prevent environment pollution. Chapter 4 discusses the shift to all electrical actuators for subsea valves.

Table 3.4 shows the actuator selection criteria for ball valves in one offshore topside project, performed according to the logic provided in the chart illustrated in Fig. 3.10. Fig. 3.11 illustrates a 38″ CL1500 fail-as-is ball valve with an electrical actuator that will be installed on an oil export pipeline.

Table 3.4 Actuator selection for ball valves in an offshore topside project.

Valve	Failure mode	ESD	Actuator selection
38″ CL1500ball valve	Fail-as-is	No	Electrical
38″ CL1500ball valve	Fail safe closed	Yes	Hydraulic
30″ CL1500ball valve	Fail safe closed	Yes	Hydraulic
20″ CL1500ball valve	Fail-as-is	No	Electrical
20″ CL1500ball valve	Fail safe closed	Yes	Hydraulic
10″ CL300ball valve	Fail safe closed	Yes	Pneumatic
8″ CL300ball valve	Fail safe open	Yes	Pneumatic
2″ CL300ball valve	Fail safe closed	Yes	Pneumatic
2″ CL600ball valve	Fail-as-is	No	Electrical

FIG. 3.11

A 38″ CL1500 ball valve with an electrical actuator.

Courtesy: Elsevier.

As mentioned above, butterfly valves require less torque than ball valves due to their low mass closure member disk and low pressure class. Therefore butterfly valves do not normally require the high torque provided by hydraulic actuators. Table 3.5 indicates the choices of actuators for some of the butterfly valves in one offshore, topside project. Fig. 3.12 illustrates the last item in Table 3.5, which is a 4″ CL150 butterfly valve with an electrical actuator undergoing a pressure test as part of a factory acceptance test.

Actuator selection to minimize emission for subsea oil and gas applications involves shifting completely from hydraulic to electrical actuators. Therefore all of the hydraulic subsea actuators listed in Table 3.3 should be converted to electrical subsea actuators. However, as it will be discussed in Chapter 4, maintaining the fail-safe mode of operation for subsea electrical actuators could be a challenge, but it is possible to be maintained.

Table 3.5 Actuator selection for butterfly valves in an offshore, topside project.

Valve	Failure mode	ESD	Actuator selection
38″ CL150butterfly valve	Fail-as-is	No	Electrical
38″ CL150butterfly valve	Fail-as-is	No	Electrical
18″ CL150butterfly valve	Fail-as-is	No	Electrical
14″ CL150butterfly valve	Fail-as-is	No	Electrical
10″ CL150butterfly valve	Fail safe closed	No	Pneumatic
8″ CL150butterfly valve	Fail safe closed	No	Pneumatic
6″ CL150butterfly valve	Fail safe closed	No	Pneumatic
6″ CL150butterfly valve	Fail-as-is	No	Electrical
4″ CL150butterfly valve	Fail-as-is	No	Electrical

FIG. 3.12

A 4″ CL150 butterfly valve with an electrical actuator undergoing a pressure test.

Questions and answers

1. Unlike valves, piping and flanges, actuators are not considered a source of fugitive emission in refineries or other oil and gas plants. Why?
   A. Actuators use air, hydraulic oil or electricity as their main source of power in most cases; these are not considered a source of fugitive emission.
   B. The number of actuators in refineries is much lower than the number of valves.
   C. Shifting to all electrical actuators in refineries reduces emission to zero.
   D. Actuators do not typically work with high-pressure fluid.

Answer: Option A is correct, since air, hydraulic oil and electricity are not sources of fugitive emission. The number of actuators in refineries is lower than the number of valves, but the number of actuators in refineries is not a reason why actuators are not sources of fugitive emission. Thus option B is wrong. Option C is not correct because shifting to all electrical actuators in refineries is not practical. There are limitations associated with electrical actuators, such as the speed of operation and the difficulty of maintaining failsafe position, so using hydraulic actuators could be necessary in some cases. Option D is totally wrong, since actuators work with high-pressure air and hydraulic oil.

2. In which case could an electrical actuator be selected to maintain an environmentally friendly solution and zero emission?
   A. A 30″ ball valve connected to a process shut down system.
   B. A 28″ butterfly valve that requires an actuator solely for ease of operation.
   C. A 6″ ball valve connected to an emergency shut down system.
   D. A 40″ CL1500 ball valve with fail safe closed function.

Answer: The ball valve in option A is connected to a process shut down system, so it should maintain a fail safe position, which is difficult to achieve with an electrical actuator. Thus option A is not correct. Option B requires an electrical actuator, since the valve is actuated just for ease of operation. An electrical actuator has a motor that can be upgraded with a gear box that can be installed on the top of the valve to produce a higher torque value. Option C is not correct, because a 6″ ball valve connected to an emergency shut down system should be fail safe closed, which cannot be accomplished with an electrical actuator. The valve in option D requires a hydraulic actuator, since it is a large, high-pressure class ball valve, so option D is not correct. Fig. 3.13 illustrates the 28″ butterfly valve in pressure class 150 with an electrical actuator for ease of operation.

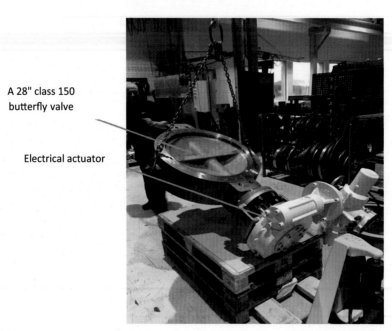

A 28" class 150 butterfly valve

Electrical actuator

FIG. 3.13

A 28″ CL150 butterfly valve with an electrical actuator for ease of operation.

3. Which sentence is not correct?

    A.   Pneumatic actuators are not common in the subsea sector of the oil and gas industry.

    B.   Electrical actuators for subsea use can maintain a fail safe closed position through a spring.

    C.   Single acting hydraulic actuators produce higher torque compared to double acting actuators with the same source of power.

    D.   A pneumatic actuator is a correct choice of actuator for a 2″ CL1500 ball valve with a failsafe closed mode of operation.

Answer: Option A is correct, since pneumatic actuators are not common for the subsea oil and gas industry. Option B is correct, since electrical actuators in the subsea sector can maintain a fail safe mode of operation. Option C is wrong, since double acting actuators can typically provide higher torque compared to single acting actuators with spring force. Option D is correct; the valve should have a fail safe closed mode of operation and it does not require high torque in such a small size, so a pneumatic actuator is the correct type. Fig. 3.14 illustrates a 2″ CL1500 ball valve with a pneumatic actuator during transportation.

4. A 12″ valve is required to be opened in 10 s. Which sentences are completely correct?

1. The required speed of valve operation is high, so most likely an electrical actuator is not a correct choice.

2. An electrical actuator is the best choice for this valve, since it provides an environmentally friendly solution.

Pneumatic actuator

Ball valve

FIG. 3.14

A 2″ CL1500 ball valve with a pneumatic actuator.

3. A hydraulic actuator is the best choice since it provides the fastest speed of operation.

4. t is not possible to select the best choice of actuator, since not all the information is available.
- **A.** 1 & 2.
- **B.** 2 & 4.
- **C.** 1 & 4.
- **D.** 2 & 3.

Answer: The valve has a very fast speed of operation, faster than one inch per second, so an electrical actuator is not suitable for this application. Thus the first choice is correct and the second choice is not correct. The third choice is not completely correct, since the pressure class of the valve is not given. The higher the pressure class is, the greater the likelihood of the suitability of a hydraulic actuator for this valve and the provided application. Option 4 is correct, as it is not possible to select the best choice of actuator with 100% accuracy without additional information. The best choice of actuator, based on the given information, is either a hydraulic or pneumatic actuator. Thus option C is the correct answer.

5. Which sentence is correct about actuator selection for a clean environment?
- **A.** Electrical actuators should be selected for all the valves inside refineries.
- **B.** Usage of hydraulic actuators should be prohibited.
- **C.** Using electrical actuators for subsea systems is an ongoing process and is the future of subsea development.
- **D.** Reducing the number of actuators in plants is recommended.

Answer: Option A is not correct, because electrical actuators have some limitations, it is not possible to select electrical actuators for all the valves in refineries. Option B is not correct, since in some cases where high torque and/or high speed of operation are required, hydraulic actuators are the best choice of actuation. Option C is correct, because the usage of all electrical actuators subsea, rather than hydraulic actuators, to protect the environment is becoming more popular and will be the future of the subsea oil and gas industry. Option D is not correct, because actuators should be used when they are required in applications such as the operation of valves in remote areas and those that need frequent and/or easy, fast and automatic operation, etc.

Further reading

Mokveld, Reducing (Fugitive) Emissions With Electric Actuators, 2020 (online) available at: Reducing (fugitive) emissions with electric actuators – Mokveld.com (accessed 29 November 2020).

D. Olson, The impact of actuation developments on fugitive emissions, Valve World Mag. 21 (5) (2016) 61–63.

K. Sotoodeh, Why are butterfly valves a good alternative to ball valves for utility services in the offshore industry? Am. J. Ind. Eng. 5 (1) (2018) 36–40, https://doi.org/10.12691/ajie-5-1-6.

K. Sotoodeh, Actuator selection and sizing for valves, Springer Nature Appl. Sci. (2019), https://doi.org/10.1007/s42452-019-1248-z.

K. Sotoodeh, All electrical subsea control systems and the effects on subsea manifold valves, J. Marine Sci. Appl. (2020), https://doi.org/10.1007/s11804-020-00155-1.

Supermec, What are the Different Hazardous Area Classifications? 2020, (online) available at: https://www.supermec.com/it/blog/2019/08/27/what-are-the-different-hazardous-area-classifications. (accessed 30 November 2020).

All-electrical subsea systems and actuation

4.1 An introduction to subsea

The term 'subsea' in the context of this book refers to the sector of the offshore oil and gas industry that operates underwater in locations such as seas and oceans. Many oil and gas resources are located under the water in offshore areas around the world. Subsea production systems are located on the sea floor. The subsea environment is known as the harshest and most corrosive environment in which oil and gas production takes place, and subsea equipment is subjected to a high load and stress from the water. Typically, a subsea production system contains subsea wells, a wellhead and production trees, subsea manifolds, tie-ins to connect two pipes together or to connect a pipe to a structure, flowlines, jumpers and an umbilical system, as well as subsea equipment or facilities such as pumps, compressors, etc. Fig. 4.1 illustrates typical subsea field architecture. The oil and gas produced from different wells are transferred to subsea Christmas trees, also called wet trees. A Christmas tree is an assembly of piping, valves, actuators and fittings installed on a well to control the fluid, mainly through a series of valves, as illustrated in Fig. 4.2. The fluid from different subsea Christmas trees is integrated in the manifolds. Manifolds are a combination of pipes, valves and structures (see Fig. 4.3), used to gather the fluid from different wells and integrate them into one header in order to reduce the bulk of the required piping system. The connections between the Christmas trees and manifolds are made through a piece of pipe called a jumper. In general, jumpers are used to connect different subsea structures together. All of the valves and actuators are located in the manifolds, Christmas trees and subsea distribution units (SDUs). It is common to have manual valves on SDU systems, meaning that actuators are rarely used on SDUs.

An SDU (see Fig. 4.4) is a subsea structure that is used to distribute chemicals, such as mono-ethylene glycol (MEG), methanol, hydraulic fluid, wax inhibitor, etc. Hydraulic oil is required for hydraulic actuators and other components that work with hydraulic oil. MEG and methanol are injected into piping systems that contain gas in order to prevent hydrate formation in the piping system. Hydrates are ice-like crystalline solids that can form inside the gas due to high pressure and low temperature conditions. Hydrate formation can cause problems in the transition of the gas and in flow assurance. Wax can become deposited in piping that contains oil service, so wax inhibitor is transmitted into the piping system inside the manifolds or in the Christmas trees through an SDU.

FIG. 4.1

Typical subsea field architecture.

Courtesy: Elsevier.

FIG. 4.2

A Christmas tree and the installed valves.

Courtesy: Shutterstock.

FIG. 4.3

A subsea manifold including valves, piping and structure.

Courtesy: Shutterstock.

FIG. 4.4

A subsea distribution unit (SDU).

Courtesy: Shutterstock.

FIG. 4.5

Subsea umbilical.

Courtesy: Elsevier.

FIG. 4.6

Subsea umbilical termination unit.

Courtesy: Elsevier.

An umbilical (see Fig. 4.5) is a combination of hoses and cables used to transport hydraulic oil and electrical power from a topside platform or host facility to a subsea system.

Umbilicals end at a Subsea Umbilical Termination Assembly (SUTA), as illustrated in Fig. 4.6. Electrical and hydraulic power are transferred between a SUTA and an SDU through hydraulic and electrical flying leads, as shown in Fig. 4.7, which illustrates EFL (electrical flying leads) and HFL (hydraulic flying leads). Hydraulic and electrical power and chemicals are transmitted from an SDU to the trees and manifolds. EFLs and HFLs are used between the manifolds, trees and SDUs to transmit hydraulic oil and electricity.

A subsea control system is a key part of a subsea development system, and is covered by ISO 13628-6, titled "Petroleum and natural gas industries—Design

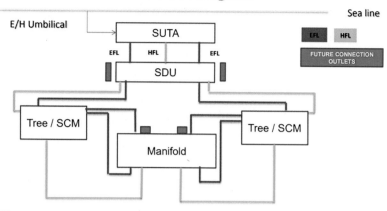

FIG. 4.7

SDU block diagram.

and operation of subsea production systems—part 6: Subsea production control system." In fact, the reliability of a subsea system during its design life, which could be as long as 30 years, is largely dependent on the proper functioning of the subsea control system. Subsea valves and actuators require electrical and hydraulic power for safe and accurate operation and functionality. The important factor in the supply of hydraulic fluid and electricity is to control their supply to the required subsea facilities, which is done through a subsea control system. The control system could be installed on a topside platform as a host facility or subsea.

Subsea manifolds may be used for pipe-to-pipe connection or tie-in. These kinds of manifolds can be a pipeline end manifold (PLEM) or a pipeline end terminal (PLET). A PLEM is a subsea structure placed at the end of a pipeline to connect a rigid pipeline with another structure or with additional piping. A PLET is used for making a single pipe-to-pipe connection as illustrated in Fig. 4.8. As an example, as illustrated in Fig. 4.1, a PLET is used to connect a jumper to a flowline. A flowline, also called a pipeline, is defined as a segment of piping that transfers the fluid coming out from the wellhead to the topside facilities from the manifolds. It is important to bear in mind that a flowline is not directly connected to a platform. Typically, a riser, which is a piece of pipe, is installed between the platform and the flowline as illustrated in Fig. 4.9. Fig. 4.1 illustrates a flow line from a PLET type manifold to a platform (host facility).

The subsea oil and gas industry is continuously moving toward more simple design to save weight and space, implement digitalization, and reduce both the expenditure (CAPEX) and operational costs (OPEX). Moreover, keeping the environment clean by preventing emission and spillage is always a concern. It is important to bear in mind that the venture for oil and gas exploration and production is

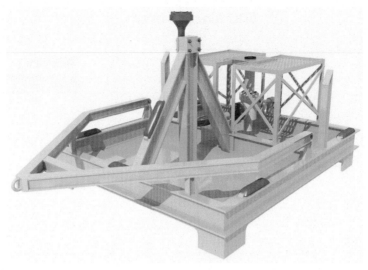

FIG. 4.8

A PLET.

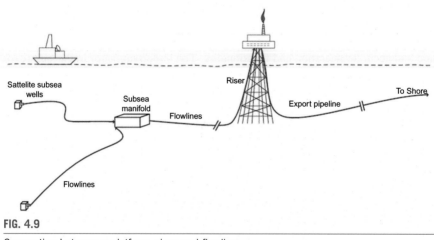

FIG. 4.9

Connection between a platform, riser and flowline.

Courtesy: Elsevier.

moving toward ever deeper and more remote areas of the seas and oceans. To achieve the desired aims of costreduction and maintaining a clean environment, an all-electrical control and actuation system is preferred over electro-hydraulic subsea distribution and control with hydraulic actuators. The use of all-electrical actuators instead of electro-hydraulic actuators has many benefits; these benefits are discussed in the following sections.

4.2 Benefits of an all-electrical subsea system

The oil industry has been keen to move to the concept of all-electrical subsea systems for many years. Many research programs have been conducted to justify and evaluate all-electrical subsea systems. Shifting from the current electro-hydraulic subsea distribution and control systems explained in the previous section to an all-electrical system has many advantages, which can be summarized as reducing cost, improving health, safety and environmental (HSE) protection, and increasing functionality, reliability and flexibility. The main aim of this section is to explain the benefits that can be achieved by changing the technology from an electro-hydraulic to an all-electrical system, which necessitates changing all the subsea hydraulic actuators to electrical types.

4.2.1 Cost

4.2.1.1 Capex

Fig. 4.10 illustrates the topside and subsea units involved in the generation and distribution of hydraulic and electric power to subsea system and facilities. The main CAPEX saving arising from the shift to an all-electrical system is related to the elimination of hydraulic fluid and the components that handle it.

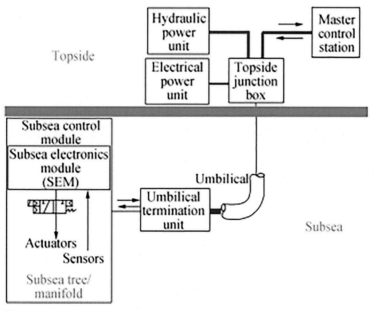

FIG. 4.10

Subsea control and distribution system.

Courtesy: Elsevier.

The components that handle hydraulic fluid include the hydraulic control system, consisting of a hydraulic power unit, accumulators, a hydraulic umbilical, and tubes and hydraulic actuators. Fig. 4.10 illustrates the pathways of hydraulic power distribution from a hydraulic power unit (HPU) to a subsea control module, which contains an accumulator and direct control valves (DCVs), and eventually to the subsea hydraulic actuators. All of the components illustrated in the figure, except the valves, are required to be removed in the course of the pivot to an all-electrical installation subsea.

The cost of electricity for both hydro-electrical and all-electrical subsea systems are almost the same, since both systems use electricity as a source of power. In fact, electrical power is required for subsea electronics and sensors, whereas hydraulic power is required for the actuation of subsea valves in electro-hydraulic subsea technology. Shifting to all-electrical actuators involves keeping and upgrading the electrical system and eliminating the hydraulic system completely. Fig. 4.11 illustrates that shifting from a conventional electro-hydraulic system to an all-electrical system for subsea manifolds results in far fewer distribution connections and components.

The estimated hydraulic consumption for a subsea unit with four wells and one manifold is five tonnes per year. The cost of using a hydraulic system for subsea operation is not limited to the direct cost of hydraulic fluid, however. Other costs associated with hydraulic systems include transportation to the offshore platform; storage, which varies due to weight and space; handling costs; and costs related to the components of hydraulic related systems such as pumps and filters located in the HPU, accumulators, etc. (Fig. 4.12).

Experiences gained from different subsea oilfield developments show that the cost reduction of using an all-electrical compared to an electro-hydraulic system is between approximately 10 and 30%. It can be concluded that removing the

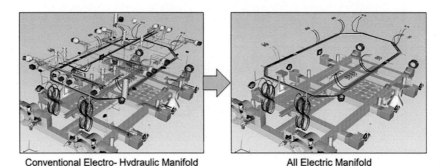

Conventional Electro- Hydraulic Manifold   All Electric Manifold

FIG. 4.11

Comparison between distribution system and components in conventional electro-hydraulic and all-electrical manifolds.

Courtesy: Offshore technology.

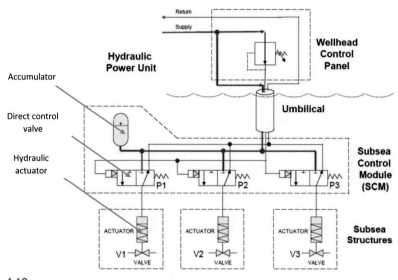

FIG. 4.12

Subsea hydraulic distribution.

Courtesy: Elsevier.

hydraulic system and components has the following effects, which entail cost reduction consequently:

- less complex system;
- less occupied space and weight;
- less cost of equipment, hydraulic fluid handling, etc.;
- less possibility of failure;
- less required equipment for installation and management.

One major saving is due to the fact that a more simple, compact and less costly *umbilical* is required for an all-electrical subsea system compared to a hydro-electrical system. The umbilical lines used for the transmission of both high and low hydraulic pressure can be eliminated due to the usage of an all-electrical subsea system. Some hydraulic systems contain a return line to the surface to avoid dumping the hydraulic fluid used inside the actuators to the sea which is called closed loop hydraulic circuit system. A closed loop hydraulic system as an environmentally friendly solution is explained in Chapter 5. The hydraulic return line and umbilical can be eliminated when an all-electrical subsea system is implemented. The list of hydraulic and electrical facilities which require to be eliminated or upgraded due to shifting to all-electrical subsea system are listed below:

A *hydraulic power unit (HPU),* sometimes called a hydraulic power pack, is a system that contains a motor, a fluid reservoir and a pump. An HPU stores hydraulic fluid and transports it to the required destination. An HPU to transmit hydraulic fluid from topside to subsea is normally installed on a host facility, such as a platform or ship. If an all-electric system is used, an HPU is no longer required on topside facilities. Cost saving thus follows from HPU elimination due to the direct cost of the unit, as well as other associated costs such as installation, commissioning and connections.

Hydraulic accumulators, also known as subsea accumulator modules (SAM), which could be installed subsea, are no longer required in a subsea all-electrical system.

Subsea control module (SCM): In total, SCM due to having all-electrical system is cheaper since no hydraulic circuit is required. But more electrical and external interfaces are required for an SCM to support electrical actuators.

The Electrical power unit (EPU) will probably need to be upgraded due to the usage of all-electrical actuators, which can increase the initial cost. In addition, *electrical actuators* in general are more expensive than hydraulic actuators. Thus, the possibility of requiring an EPU upgrade and the need to purchase electrical actuators are two factors that could increase the CAPEX for an all-electrical subsea system.

4.2.1.2 OPEX

Reducing OPEX as a result of transitioning to electrical actuator usage in subsea fields can be discussed from four different angles. The first is related to *reliability and lower operational costs associated with electrical actuators,* which is explained in more detail in the subsection titled "reliability." The second parameter in reducing OPEX is *elimination of hydraulic oil consumption.* Dealing with hydraulic fluid entails the extra costs of transporting it to the offshore facilities, storing and handling it, and incurring the operational costs associated with the pumps, filters and other equipment that contain and move the fluid. The third parameter is *process optimization and efficiency,* which is connected to the more accurate control of choke valve positioning and flow control after the wellhead obtained by using electrical actuators. No cost model has been developed so far to evaluate OPEX reduction due to the accuracy of the electrical actuators installed on choke valves. But generally speaking, hydraulic actuators may open the choke valves less than they should be, which leads to loss of production. On the other hand, opening a choke valve more than it should be could create operational problems due to high flow and associated costs. The fourth angel is related to *less production loss* due to using the electrical actuators.

4.2.2 HSE

Annually, a large volume of hydraulic fluid is drained into the environment during normal operation due to both unplanned leakage and the usage of open loop hydraulic systems. The main characteristic of an open loop hydraulic system that the hydraulic fluid is discharged into the marine environment during the actuation of the subsea valves. Both open and closed loop hydraulic systems are explained in

Chapter 5. An all-electrical solution is therefore an attractive solution in terms of HSE issues. The obvious HSE benefit derives from the removal of all the hydraulic fluid, thereby eliminating the chance of a hydraulic oil spill to the subsea environment and preventing the exposure of personnel to the hydraulic fluid.

The possible contamination caused by hydraulic fluid is not limited to subsea. Hydraulic systems must be tested and transported onshore, so all the onshore handling and testing areas should be considered as possible locations that could be polluted by hydraulic oil. With no hydraulic system, there is no high pressure in the actuators, which improves the safety of the personnel who are working with and handling the subsea actuators. Thus, in conclusion, three elements: no fluid discharge, no hydraulic exposure to personnel, and no high pressure are the main areas of HSE improvement that can be obtained by using an all-electrical instead of an electro-hydraulic subsea system.

4.2.3 Reliability

All-electrical system and actuation reliability is reviewed in this section in different aspects, such as *very good operator companies' experience with subsea electrical actuators*, *ease of maintenance and replacement* and the possibility of *monitoring electrical actuators online from the control room*.

This section explains in more details about very good experience of operator companies about using the subsea electrical actuators for almost twenty years. The development of subsea electrical actuators began in 1999 and the first electrical actuators were installed in a subsea field in 2001 on nonsafety critical valves. Safety critical valves include but are not limited to those used in emergency shut down (ESD) and high integrity pressure protection systems (HIPPS). Some of the main end users or operators in the oil and gas industry have claimed to have very good experiences with electrical actuators. Schlumberger, the largest oilfield service company, has logged more than one million hours of using subsea electrical systems. These systems are designed to operate for 25 years without failure with a 20% cost reduction compared to electro-hydraulic control systems. Equinor, the major oil company and end user on the Norwegian continental shelf, has been using subsea electrical actuators with a fail-as-is mode of operation for more than 15 years and have noted very good operational experience. It is important to bear in mind that Equinor launched the first subsea all-electrical system in the world. In 2017, Equinor announced the successful performance of electrical actuators for over 700,000 h. Having established the reliability of all-electrical operation in its production systems, Equinor has continued to shift to all-electrical systems, including electrical actuators.

While Equinor successfully operated all types of valves with electrical power, including those located on Christmas trees, manifolds and subsea distribution units, the company has preferred to operate the final valve, called the surface controlled subsurface safety valve (SCSSV), with a hydraulic system. A SCSSV valve is an emergency shut down valve installed in the well to shut down the well in an emergency situation to prevent catastrophic blowout. A blowout is defined as an uncontrolled release of crude oil and/or gas from a well. Modern wells have a blow-out preventor as well as an SCSSV. Operation of SCSSV is more reliable with hydraulic fluid than an electrical power.

Some of the main engineering, procurement and construction (EPC) contractors in subsea business, such as AkerSolutions, Baker Hughes and Technip FMC, are producing subsea electrical actuators. FMC technologies has logged more than eight million operating hours of electrical actuator usage without any significant reported issue. Other reputable valve and actuator manufacturers, such as Advanced Technology Valve (ATV) and Petrol Valve (PV), are producers of subsea electrical actuators. The next two paragraphs compare reliability of electrical and hydraulic actuators from possibility of failure and maintenance point of view.

Hydraulic actuators have few parts, which reduce their possibility of failure and improve their reliability. However, potential issues include the contamination of the hydraulic fluid and the possibility of hydraulic leakage from the components handling the hydraulic fluid. Contamination of hydraulic fluid with particles could damage the actuators and components that handle the hydraulic fluid, such as the directional control valve (DCV) and the subsea control module (SCM). A problem with DCVs may require changing the whole subsea control module, which results in production stop. Stoppage of production means losing money and increasing OPEX. Additionally, failure of hydraulic actuators is costly, since hydraulic actuators are not retrievable or replaceable in most cases. Fig. 4.13 illustrates a subsea control module that is a part of a subsea control system. It includes two DCVs for changing the direction of the hydraulic supply to the actuator or altering the direction of hydraulic return from the actuator.

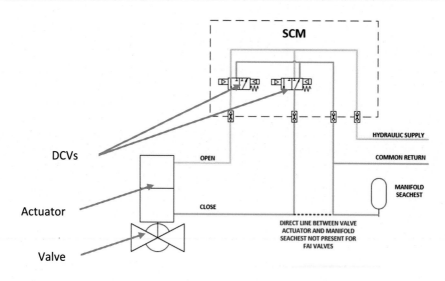

FIG. 4.13

Subsea control module with directional control valves connected to hydraulic actuated valve.

Table 4.1 AET vs. EHT reliability comparison.

Production percentage	Number of active wells	AET availability percentage	EHT availability percentage
100	4	86.5	84.3
>75	3 or 4	96.9	96.4
>50	2, 3 or 4	97.5	97.0
>25	1, 2, 3 or 4	97.5	97.0

Electrical actuators have many parts and sensitive components, and are thus more complex than hydraulic actuators, but they have dual redundance to a large extent. Also, it is easy to replace electrical actuators underwater with the aid of a remote operated vehicle (ROV). In most cases, production can continue during the replacement of electrical actuators. Another advantage of electrical actuators that improves reliability is the possibility of constantly monitoring their status from the control room. A joint study between British Petroleum (BP) and Cameron, two different end users in the oil and gas industry, and an engineering group at Cranfield University was organized in 2004 to evaluate and compare the reliability of All-Electrical Technology (AET) vs. Electro-Hydraulic Technology (EHT). The result of this reliability evaluation is provided in Table 4.1.

4.2.4 Flexibility and efficiency

Electricity is known as a flexible source of energy, which means that it has many more uses compared to oil. Many different types of household equipment, such as heaters, light bulbs, etc., use electricity. Electricity can even be used in vehicles as a source of energy. Electricity is also flexible in the sense that it does not need to be separated from other products like crude oil does. Crude oil must be separated into a variety of useful products through a refining process. In general, the usage of electrical power is growing very quickly worldwide. The other aspect is flexibility in the context of using all-electrical systems is that expanding a field or adding new equipment to a field that works with an all-electrical system is much easier than expanding or adding to a field that works with a hydro-electrical system.

One of the limitations or challenges of electro-hydraulic systems is evident when the subsea field is located in a very deep area and the pressure of the reservoir and produced oil is extremely high. In such cases, the system must be covered by valves in 20,000 psi according to API 17D/ISO 13628-4, which are standards for subsea wellheads and tree equipment. In such instances, the combination of high internal pressure inside the actuated valve and the high head of the seawater at deep depth results in the necessity of using an extremely high-pressure API class valve. The valve must have thick walls and heavy internals. Therefore a very high pressure of hydraulic fluid and/or a very large hydraulic actuator is required for moving such a valve. Some components of the hydraulic transmission system, such as pumps,

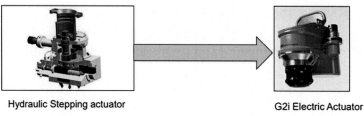

Hydraulic Stepping actuator   G2i Electric Actuator

FIG. 4.14

Changing a hydraulic stepping actuator to an electrical actuator.

Courtesy: Offshore technology.

accumulators and the hydraulic supply umbilical must also be larger and heavier to some extent. Pumping the hydraulic fluid to such a depth with large pressure is impractical and thus not a good engineering practice when it comes to energy efficiency. Shifting to electrical actuators will solve this challenge, and is an efficient and logical engineering decision.

4.2.5 Functionality

Electrical power has a faster speed of transmission compared to hydraulic power, which means that electrical actuators provide faster response time and operation compared to hydraulic actuators. In addition to quick operation, electrical actuators have higher accuracy compared to hydraulic. Fig. 4.14 illustrates that replacing a hydraulic stepping actuator with an electrical actuator on a choke valve results in faster and more accurate operation, less vibration, more accurate positioning and easier maintenance and retrievability of the actuator, while eliminating a large amount of hydraulic consumption. A subsea choke is the preliminary control valve installed after a Christmas tree to reduce the pressure and control the flow of the produced fluid from the wellhead.

Table 4.2 provides a summary of comparison between hydraulic and electrical actuator systems.

Table 4.3 summarizes the advantages and disadvantages of each type of system and actuation.

4.3 Subsea valves and applications

The main aim of this section is to discuss the type of subsea valves which are working with subsea actuators as well as their locations and applications. Generally, four types of valves are actuated in subsea environments. Through conduit gate (TCG) valves, choke valves, ball valves and safety valves such as subsurface safety valves. In general, all these valves except SCSSVs are compatible with electrical actuators.

Table 4.2 Comparison between hydraulic and electrical actuators.

Features and differences	Electrical actuator	Hydraulic actuator
Energy consumption	Electrical power which is lower compared to hydraulic	Hydraulic oil
Environmentally friendly	More environmentally friendly	Less environmentally friendly
Space consumption	More compact	Less compact
Risk of leakage	No	Yes
Fail safe option	Yes, by using spring force and loss of power	Yes, by using spring force and loss of hydraulic
Reliability	More	Less
Flexibility of usage in different locations	Yes	Yes

Courtesy: Springer.

Table 4.3 Advantages and disadvantages of subsea electrical and hydraulic actuators.

	Advantages	Disadvantages
Hydraulic actuators	- Provide a large amount of torque or force of actuation due to the high pressure of oil. - Achieving a fail-safe mode during operation is easy by means of spring force.	- Hydraulic actuators are used in electro-hydraulic systems with two types of systems: hydraulic and electrical. This increases the cost and adds complexity to the subsea development. - Hydraulic actuators, especially rotary actuators like rack and pinion, are complex and relatively large compared to electrical actuators. - Issue with lack of safety and exposure of personnel to high-pressure hydraulic fluid - Possibility of spill to the environment - Higher OPEX and CAPEX
Electrical actuators	- Suitable for both linear and rotary movements of the valves. - High speed and high accuracy - High reliability with the possibility of self-diagnosis and online monitoring - Lower cost, simpler and more functional	- Maintaining fail-safe mode in the event of power loss is challenging in some cases. - Linear movement requires a transition system.

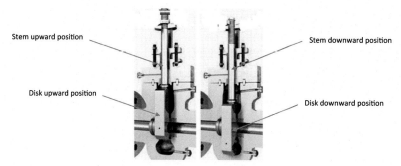

FIG. 4.15

Through conduit slab gate valve with linear movement of stem and disk.

Historically, API 6A TCG valves have been used for subsea field development, especially for Christmas trees and wellheads. TCG valves have robust design and are suitable for particle- containing or dirty fluid services. Gate valves are used for wellheads and Christmas trees; these are typically TCG valves from 1 13/16″ to 7 1/16″ in size. TCG valves could be slab, single or double expanding types. The type of valve stem and internal movements are linear, as illustrated in Fig. 4.15.

Choke valves used after Christmas trees to reduce the pressure of the fluid and provide fluid flow control have a stem and valve internal with linear movement in most cases, like TCG valves. Conversely, ball valves have a 90-degree rotation of their ball (closure member) and stem; this type of valve and valve motion is called "quarter turn" (see Fig. 4.16). As ball valves do not have linear motion of their stem and closure member, the conversion of rotational to linear movement is not applicable for ball valves.

Fig. 4.17 illustrates a typical process flow diagram for a subsea manifold. The manifold is connected to six wells, so it is called a six-slot manifold. All of the connections coming from the six wells initially flow into two headers, and then into one header in the manifold. The header in the manifold is connected to flowline A from one side and to flowline B from the other side. The valves located on the header of the manifold are ball valves, and the small valves located on the manifold branches are TCG valves. The gate valves on the branches are normally open and get closed to isolate the manifold from the wellhead in the event of operational problems in the wells. The Christmas tree valves, shown in Fig. 4.2, are typically TCG valves.

4.4 Subsea electrical actuator design review

The benefits of all-electrical subsea systems and electrical actuators were discussed in the previous section. This section takes a closer look at the design of subsea electrical actuators. In general, three essential parts are used in subsea electrical actuators: the *gear*, *motor* and *motor controller*. Four important design aspects should be

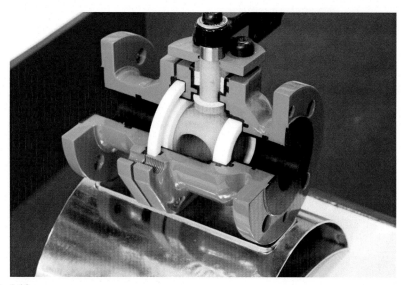

FIG. 4.16

A ball valve.

Courtesy: Shutterstock.

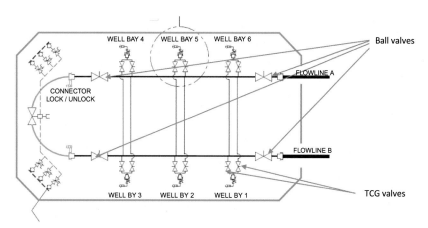

FIG. 4.17

Six-slot manifold process flow diagram.

considered for subsea electrical actuators: *packaging the motor drivers, failsafe solution, redundancy* and *conversion of rotational to linear movement*. It should be noted that the conversation of rotational to linear movement is applicable to through conduit and choke valves most of the time. Packaging of electrical actuators is covered by ISO 13628-6 standard.

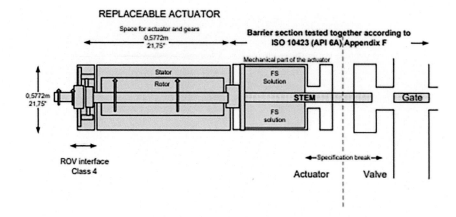

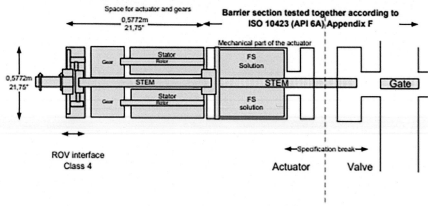

FIG. 4.18

Direct drive electrical actuator (top) and electrical actuator with small motor and gear (bottom).

4.4.1 Direct drive vs. gear-operated motors

Two electrical actuator design types are reviewed in this chapter. The first is direct drive design, in which just a motor without any gear is used for moving the valve (see Fig. 4.18, top section). The second is a small motor with gearing (see Fig. 4.18, bottom section). Direct drive motors are typically larger motors that can produce sufficient force or torque for valve operation without any need for a gearbox. In a direct drive electrical actuator, a magnetic field is created in the stator, which is a stationary part of the electrical motor or actuator. The magnetic field makes the rotor rotate. The rotary motion of the rotor is transferred to the extended part of the stem in the actuator and creates the rotary movement of the stem. The rotary motion of the stem is transferred to linear movement, since a gate valve is typically connected to the electrical actuator through a screw mechanism, explained later in this chapter. The second design involves using a gear box inside the electrical actuator to increase the actuator torque or force. A gear box is required for relatively smaller motors. The gear box is driven by the motor, including the stator and rotor, which are located around the stem. The gear rotation is transferred to the stem consequently.

4.4.2 Fail-safe mode of operation

The type of electrical actuator considered in this section is a retrievable actuator with the possibility of replacement and maintenance to provide flexibility in operation. If the electrical actuator does not have a fail-safe function, then the design could be more compact; note that the area at the bottom part of left side of the actuator in Fig. 4.18, indicating failsafe (FS) solution, could be eliminated. If a fail-safe function is required, then two choices are typically available for electrical actuators: using a mechanical spring or using an electro-mechanical spring, also called a battery. The two choices of spring for electrical actuators are illustrated in Fig. 4.19. The use of a spring as a mechanical device for closing a fail-safe close valve or opening a fail-safe open valve is widely practiced for hydraulic actuators in subsea and other sectors of the oil and gas industry. The problem with a spring is that its size must be increased significantly to provide higher force for closing or opening a high-pressure class valve installed at a deep water depth. Therefore electro-mechanical devices such as an electromagnet or a battery can be used as alternatives to a spring.

4.4.3 ROV override

The other important feature of electrical actuators is that they are equipped with ROV override. ROV override allows the valve to be cycled (opened and closed) independently of the actuator. ROV override class 4 can be selected for the electrical actuator as an option. The ROV interface and classes are typically selected according to

 VS.

FIG. 4.19

Mechanical vs. electro-mechanical spring for fail-safe closed subsea electrical actuators.

Table 4.4 ROV classes and maximum torque capacity as per API 17H/ISO 13628-8.

Class	Maximum ROV bucket design or input torque, N m (lbf; ft)
1	67 (50)
2	271 (200)
3	1355 (1000)
4	2711 (2000)
5	6779 (5000)
6	13,558 (10,000)
7	33,895 (25,000)

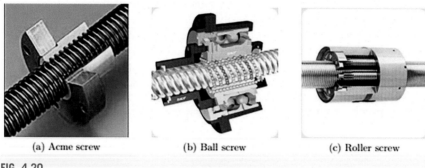

    (a) Acme screw   (b) Ball screw   (c) Roller screw

FIG. 4.20

Screw types for transmission of rotary motion to linear motion.

ISO 13628-8 or API 17H, titled "Remotely Operated Vehicle (ROV) on subsea production systems." Seven classes of ROV are defined in API 17H/ISO 13628-8, and each class has a maximum design torque capacity that is defined in Table 4.4. The higher the class number, the greater the torque input capacity.

ROV bucket classes 2, 3, and 4 are common for subsea valves. Class 3 is less popular than the other two classes for the ROV bucket. In addition, a low torque paddle could be used for smaller valves such as 0.5″ ball valves in 10,000 psi pressure class.

4.4.4 Conversion of rotation to linear movement

A common method for translating rotational movement into linear movement is the use of a screw mechanism. Fig. 4.20 illustrates different screw mechanisms common for the purpose.

The first picture on the left illustrates an Acme screw, a type used in the oil and gas industry for many years. But the main problem with the Acme screw is that it has very low efficiency. The efficiency of Acme screws could be between 20% and 80%. The type of material and lubrication affect the efficiency of Acme screws.

Table 4.5 Comparison of screws for converting rotary to linear movement.

	Roller screw	**Acme screw**	**Ball screw**
Load rating	Very high	High	High
Self-locking	No	Yes	No
Efficiency	Relatively high	Relatively low	Relatively high
Speed	Very high	Low	Moderate
Acceleration	Very high	Low	Moderate
Stiffness	Very high	Very high	Moderate
Shock load	Very high	Very high	Moderate
Space requirement	Minimum	Moderate	Moderate
Friction	Low	High	Low
Maintenance	Low	Low	Moderate
Life time	Very long	Low to moderate	Moderate

Courtesy: NTNU.

The efficiency of a screw in this context is defined as how well it can convert rotary movement to linear movement. There are some advantages related to Acme screws, which are mentioned as follows:

- Self-locking capability: When a screw is not under power, the load can push the screw and rotate it. The Acme screw has self-locking capability, which can secure its position against the load.
- Low cost: Compared to ball screws, Acme screws are 25%–80% cheaper.
- Robust design and less maintenance are required for Acme screws.

Due to the low efficiency of Acme screws, other type of screws could be selected, such as ball and roller screws. The disadvantages of both ball and roller screws are their inability to self-lock. Thus for a fail-safe spring return electrical actuator, a kind of stoppage mechanism for the spring is required when ball or roller screws are used. Table 4.5 compares all of the properties of these types of screws.

4.5 All-electrical subsea actuators: A case study for manifolds

The main aim of this section is to provide an analysis of changing the hydraulic actuators for the actuated valves located on subsea manifolds to electrical actuators. Three main questions are answered in this section:

- What types of electrical actuators are required for manifold valves? One choice is spring return fail safe mode and the other one is fail-as-is without any spring.
- Does using the electrical actuators change the requirements for safety integrity level (SIL) and associated certification?
- Does the implementation of electrical actuators on the valves change the valve selection philosophy and the type of the actuated valves?

Nine subsea projects in different parts of the world, such as Africa and Australia, have been selected for this case study of the transition to all-electrical actuators. The first finding is that only 40 actuated valves on these nine projects are actuated. The selected valves are either slab type TCG valves or ball valves, which are all located on the subsea manifolds. The size of the valves ranges from ¾″ to 18″ and the pressure classes could be either 5000 psi, 7500 psi or 10,000 psi. Twenty-nine valves out of the 40 valves have a fail-safe closed function while just eight valves have a fail-safe open function or mode of operation. The three remaining valves, including one slab gate valve and two ball valves, have a fail-as-is function or mode of operation. The result of this analysis is provided in Table 4.6.

The first conclusion based on the data provided in the table is that the slab gate valves mainly have a fail-safe close function, whereas ball valves could have both a fail-safe closed and open condition. In fact, slab gate valves, which are typically installed on the branches of the manifolds, are normally open and get closed in the event of losing hydraulic power. Slab gate valves get closed in case of failure to protect the manifolds if something goes wrong on the upstream side of the manifold on the wellheads, inside the well or on the Christmas trees. The analysis shows that 37 of the actuators have a fail-safe mode and only three electrical actuators have a fail-as-is function. Therefore one of the main challenges of electrical actuation for subsea manifolds is that approximately 97% of the actuated valves based on this analysis require fail-safe electrical actuators, which is possible to maintain—but challenging.

Safety integrity level (SIL) is a part of an international standard such as IEC 61508 that provides suppliers and end users with a common framework to design products and systems for safety-related applications. SIL provides a scientific and numeric approach to specifying and designing safety systems, enabling risk of failure to be quantified. IEC 61508 is an international standard published by the International Electro-technical Commission that addresses the functional safety of electrical, electronic and programmable electronic safety- related systems. Manifolds do

Table 4.6 Summary of analysis for usage of electrical actuators in subsea manifolds.

Manifold valves	Fail-safe close (FSC)	Fail-safe open (FSO)	Fail-as-is (FAI)
Slab gate valve	22	1	1
Ball valve	7	7	2
Total numbers	29	8	3
Type of electrical actuators	Spring return or magnetic or battery fail safe closed electrical actuator direct drive or gear motor operated	Spring return or magnetic or battery fail safe open electrical actuator direct drive or gear motor operated	No spring fail-as-is actuator

Courtesy: Springer.

not have any safety function such as emergency shut down or process shut down. In addition, high integrity pressure protection systems (HIPPS) are not applicable to manifold valves. Therefore manifold valves are not subject to safety integrity level (SIL). The Norwegian oil and gas association has produced a list of valves in different sectors of the oil and gas industry that are subject to SIL. Manifold valves are not subject to SIL, as per the data provided by the Norwegian oil and gas association. The conclusion is that the SIL requirement mainly depends on the safety functionality of the valve and not the type of actuator. Proving that an actuated valve can achieve a specific SIL through a report called a safety analysis report (SAR) is done for safety critical valves with safety function to improve safety and reliability. However, manifold valves are not safety critical valves and do not have any safety function, so the reliability of manifold valves is typically increased through more rigorous factory acceptance and qualification tests, tighter manufacturing tolerances and the use of corrosion-resistant alloys for these valves.

The last question answered in this section is about whether the valve selection philosophy for the manifolds changes due to the shift to all-electrical actuators. The general philosophy for valve selection for use in manifolds is to use TCG valves up to and including 7 1/16″ size for the branches and using ball valves for manifold headers in sizes above 8″. Using ball valves for the branches of manifolds requires a new qualification for the ball valves to validate their design for small sizes. Adding qualification means increasing the valve manufacturing and procurement time and cost, which is not desirable. The second problem with using ball valves in the branches instead of TCG valves is that ball valves are not as robust as TCG valves in particle-containing services. Historically, TCG valves have been widely used successfully for Christmas trees where the produced oil and gas plus water passing through the valves include a large amount of sand. Subsea ball valves are metal seated, and the seats of the valves could be damaged if particles get stuck between the ball and seat of the valve during opening or closing. However, the seats of through conduit gate valves are not at the risk of damage from particles. In addition, the constant operation of ball valves could lead to the wearing of the seats and possible leakage. Therefore it is not recommended, based on this author's experience, to change the manifold valve selection philosophy due to shifting from hydraulic to electrical actuators.

Questions and answers

1. Which subsea structures contain valves and actuators?
   - **A.** Jumper, flowline, subsea distribution unit (SDU).
   - **B.** Umbilical, manifold, Christmas tree.
   - **C.** Manifold, Christmas tree.
   - **D.** Umbilical termination assembly (UTA), subsea well.

Answer: Option A is not correct, since jumpers and flowlines are in fact piping components without any valves or actuators. SDUs contain subsea valves which can have actuators in rare cases, although many subsea valves on SDUs are manually operated, so it could be the case that no actuators are used for the subsea valves located on SDUs. An umbilical is a subsea component designed for electricity and hydraulic power transmission from topside to subsea through a combination of cables and tubes. Thus option B is not correct. Option C is correct, since valves and actuators are typically used inside manifolds and Christmas trees. Option D is not completely correct, since there is no valve or actuator in a UTA. However, subsea wells do contain a subsurface safety valve.

2. Which choice is correct regarding the advantages of all-electrical over electro-hydraulic systems subsea?
   A. All-electrical systems have higher CAPEX and OPEX over electro-hydraulic systems.
   B. Supplying a large amount of hydraulic fluid to remote areas in deep depths and for high- pressure class subsea valve and actuator operation is a challenge.
   C. All-electrical systems are not always as reliable as electro-hydraulic systems.
   D. Electro-hydraulic systems are not as safe as all-electrical systems due to the possibility of oil spill to the environment as well as environmental pollution.

Answer: Option A is not correct, because using an all-electrical system reduces both cost of expenditure (CAPEX) and operational cost (OPEX). Option B is correct, because a high-pressure class actuated valve located at a very deep depth requires high pressure and a large volume of hydraulic oil for successful operation. To transmit and supply such a large volume and high pressure of hydraulic fluid is a challenge due to the larger size and volume of the necessary hydraulic components, such as the umbilical. The reliability of an all-electrical system is higher than that of an electro-hydraulic system, so option C is not correct. Option D is not correct, because the lack of safety in a hydraulic or hydro-electrical system is due to dealing with high pressure hydraulic fluid and the potential exposure of personnel to the hydraulic oil. The risk of polluting the environment due to oil spill makes hydraulic or electro-hydraulic solutions less environmentally friendly compared to all-electrical systems.

3. Which sentence is not correct regarding the comparison between subsea electrical and hydraulic actuators?
   A. Maintaining a fail-safe position could be a challenge for subsea electrical actuators, unlike subsea hydraulic actuators.
   B. Maintaining a fail-safe position in subsea hydraulic actuators is typically achieved by means of a spring. However, subsea electrical actuators should be equipped either with a mechanical spring or an electro-mechanical spring to maintain a fail-safe closed position.
   C. Subsea electrical actuators have more parts compared to hydraulic ones.
   D. Self-diagnosis and online monitoring of subsea hydraulic actuators make them more reliable than subsea electrical actuators.

Answer: Options A, B and C are correct. Option D is not correct, because the possibility of online monitoring and self-diagnosis is applicable to subsea electrical actuators, not subsea hydraulic actuators.

4. Which options are correct regarding actuated subsea valves?
   A. Subsea ball valves could be actuated. Ball valves are typically installed inside manifolds.
   B. Subsea TCG valves are installed on Christmas trees for control of the fluid.
   C. Choke valves are the type of valves installed after a Christmas tree to regulate the flow.
   D. Choke valves are quarter turn like ball valves.

Answer: Option A is correct, because ball valves are typically installed inside manifolds as well as in SDU and can be actuated or manual. Option B is not correct, since TCG valves are not suitable for flow control. Using TCG valves for flow control leads to chattering, vibration and wear of the valves. Option C is correct, because choke valves are installed after Christmas trees for fluid control. Option D is not correct, since choke valves could have either linear or rotary movement depending on the arrangement of their internals.

5. Which sentence is not correct regarding cost comparison between all-electrical and electro-hydraulic subsea systems?
   A. Electrical actuators are typically cheaper than hydraulic actuators.
   B. An all-electrical system provides saving on umbilical cost compared to an electro-hydraulic system.
   C. Improving the reliability of electrical systems could save operational costs compared to using an electro-hydraulic subsea system.
   D. Elimination of hydraulic fluid from electro-hydraulic systems results in less expenditure and lower operational costs.

Answer: All the options except option A are correct. Electrical actuators are typically more expensive than hydraulic actuators.

6. Which sentence is correct regarding the use of all-electrical actuators for subsea manifold valves?
   A. Usage of electrical actuators for manifold valves necessitates the application of safety integrity level (SIL) requirements.
   B. Most subsea valves on manifolds require electrical actuators with a fail-as-is mode of operation.
   C. The valve selection philosophy for subsea manifolds could be changed due to changing the actuator from hydraulic to electrical.
   D. Actuated valves on manifolds do not have a safety shut down, process shut down or HIPPS, so SIL requirements are not applicable to these valves at all.

Answer: SIL is not applicable to manifold valves, since they are not connected to ESD or process shut down systems and they do not have HIPPS application. Thus option A is wrong and option D is correct. Option B is not correct, because most subsea manifold valves have a fail-safe function. Option C is not correct, since shifting from hydraulic to electrical actuators does not change the philosophy of valve selection for manifolds.

7. Which sentence is not correct regarding the transmission of rotary to linear movement for electrical actuators?
   A. Gate valves have linear movement, so they require a mechanism for changing the rotary motion to linear when they are electrically actuated.
   B. Roller, ball and Acme screws are three types of screws that are used in electrical actuators to change the rotary motion to linear.
   C. Fail-as-is electrical actuators require linear motion, while fail-safe-closed actuators can have rotary motion.
   D. The efficiency of a screw means the capability of transmitting the rotary force to linear force.

Answer: All options are correct except option C. Fail-as-is and fail-safe closed actuators could have either linear or rotary motion depending on the type of valve.

References

[1] D. Abicht, G.R. Halvorsen, in: Subsea all-electric, Offshore Technology Conference, Equinor, Houston, Texas, 2017.

[2] American Petroleum Institute (API) 17D, Design and Operation of Subsea Production Systems, Subsea Wellhead and Tree Equipment, second ed., API, Washington, DC, USA, 2011.

[3] American Petroleum Institute (API) RP17H, Remotely Operated Tools and Interface on Subsea Production Systems, third ed., API, Washington, DC, USA, 2019.

[4] Y. Bai, Q. Bai, Subsea Engineering Handbook, first ed., Elsevier, Atlanta, GA, USA, 2012.

[5] Helix Linear Technologies, Engineering Characteristics of a Precision Acme Screw (Online), 2020. Available at: Engineering Characteristics of a Precision Acme Screw | Helix Linear Technologies (accessed 4 December 2020).

[6] International Electrotechnical Commission (IEC), IEC 61508, Functional Safety of Electrical/Electronic/Programmable Electronic Safety-Related Systems, Part 4: Definitions and Abbreviations, second ed., International Electrotechnical Commission (IEC), Geneva, Switzerland, 2010.

[7] International Organization of Standardization (ISO) 13628-6, Petroleum and Natural Gas Industries—Design and Operation of Subsea Production Systems—Part 6: Subsea Production Control Systems, second ed., International Organization of Standardization (ISO), Geneva, Switzerland, 2006.

[8] International Organization of Standardization (ISO) 13628-8, Petroleum and Natural Gas Industries—Design and Operation of Subsea Production Systems—Part 8: Remotely Operated Vehicle (ROV) on Subsea Production Systems, first ed., International Organization of Standardization (ISO), Geneva, Switzerland, 2006.

[9] International Organization of Standardization (ISO) 13628-4, Petroleum and Natural Gas Industries—Design and Operation of Subsea Production Systems—Part 4: Subsea Wellhead and Tree Equipment, second ed., International Organization of Standardization (ISO), Geneva, Switzerland, 2010.

[10] E.W. Larssen, Design of an Electrical X-Mas Tree Gate Valve Actuator, Norwegian University of Science and Technology, Department of engineering Cybernetic, 2007.

[11] E.W. Larssen, D. Massie, K.G. Eriksson, Subsea all-electric technology: enabling next generation field developments, in: Offshore Technology Conference, AkerSolutions, Houston, Texas, 2016.

[12] T. Myhrvold, Subsea All Electric is Here To Stay (online), 2020. Available at: Subsea All Electric Is Here To Stay – DNV GL (accessed 3 December 2020).

[13] Norwegian Oil and Gas Association, Application of IEC 61508 & 61511 in the Norwegian Petroleum Industry (Recommended SIL Requirements), Revision 3, 2018.

[14] J. Praveen, M. Pathan, K. Ansari, Hyperbaric pressure testing of a subsea valve to validate deep water condition, Int. J. Mech. Prod. Eng. Res. Dev. 8 (2) (2018) 1011–1022.

[15] M. Theobald, in: Benefits of all-Electric Subsea Production Control Systems, Offshore Technology Conference, British Petroleum, Houston, Texas, 2005.

[16] K. Sotoodeh, A review on subsea process and valve technology, J. Marine Syst. Technol. 14 (2019) 210–219, https://doi.org/10.1007/s40868-019-00061-4.

[17] K. Sotoodeh, Safety integrity level in valves, J. Fail. Anal. Prevent. 18 (3) (2019), https://doi.org/10.1007/s11668-019-00666-2. ISSN 1547-7029.

[18] K. Sotoodeh, All electric subsea control system and effects on subsea manifold valves, J. Marine Sci. Appl. 19 (2020) 465–472, https://doi.org/10.1007/s11804-020-00155-1.

[19] University of Calgary, Energy Education (online), 2020. Available at: Electricity as an Energy Currency – Energy Education (accessed 3 December 2020).

Subsea hydraulic system development for reducing emission

5

5.1 Introduction

Although an all-electrical subsea system is a state-of-the-art system that is recommended and has already been applied to some subsea field developments due to its variety of advantages, there are still some subsea fields developed with electro-hydraulic systems and hydraulic actuators. Hydraulic systems provide a clean and stable supply of hydraulic fluid for subsea valves and actuators installed on the subsea bed in subsea production systems. One of the main questions in the selection of a subsea hydraulic field is whether to choose an open or a closed loop hydraulic system. The main characteristic and disadvantage of a closed loop system is that the hydraulic fluid is discharged into the sea or ocean after being used in the hydraulic actuators. In an open loop system, the hydraulic oil is transported to the topside facilities through an umbilical after being used for the actuation of hydraulic actuators. A subsea umbilical is a combination of hoses and wires used to transport hydraulic power, electricity and chemicals to subsea units. Although there are types of hydraulic fluid that could be selected for use in an open loop system to achieve a lower environmental impact, an open loop option nonetheless releases a large amount of hydraulic oil into the marine atmosphere, which is not eco-friendly. The selection of a closed loop hydraulic system is environmentally friendly, but has a higher cost due to the transmission of the hydraulic oil to the topside facilities through a return line and extra umbilical. A return line also may be called "common return line" or "hydraulic return line" is a combination of pipes that transport the hydraulic fluid used for the functioning of the actuator to the surface facilities as part of a closed loop system. In addition, using a closed loop system brings more complexity to the actuator design and qualification. Although the selection, design and installation of a subsea closed loop system is more complex, challenging and costly than that of an open loop system, the recommendation of this chapter is to select a closed loop hydraulic subsea system to reduce emissions and protect the environment. The aim of this chapter to compare open and closed loop hydraulic systems and to evaluate the effect of closed loop system selection on actuator design and qualification.

Prevention of Actuator Emissions in the Oil and Gas Industry. https://doi.org/10.1016/B978-0-323-91928-9.00005-0

5.2 Open and closed loop hydraulic systems

5.2.1 Introduction and description

The hydraulic supply to subsea actuators in electro-hydraulic subsea systems is initiated from a hydraulic power unit (HPU) in a topside facility. An HPU is a combination of a hydraulic oil reservoir or storage tank plus a pump to move the hydraulic fluid. An accumulator is a storage vessel for pressurized hydraulic oil that releases the fluid when it is required to maintain a steady flow and pressure of the pump and dampen shock and vibration. The hydraulic fluid is transmitted through an umbilical to an umbilical termination assembly (UTA) and finally to a subsea control module (SCM). Directional control valves (DCVs), located in the SCM, direct the hydraulic fluid to the failsafe closed (FSC) actuators. The difference between open and closed loop hydraulic systems begins after valve actuation, as illustrated in Fig. 5.1.

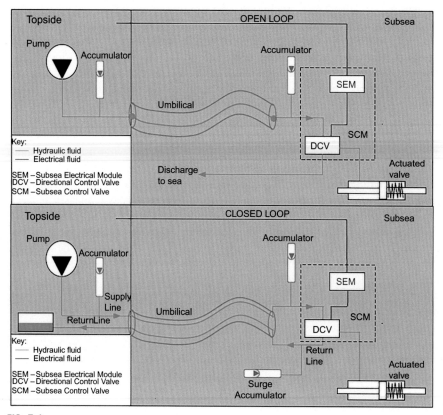

FIG. 5.1

Open versus closed loop hydraulic systems.

Courtesy: URS Cooperation Limited.

The hydraulic fluid, which enters the hydraulic cylinder through the green line, acts against the spring to open an FSC actuated valve so that the hydraulic oil moves the piston back and the spring is compressed. When the valve is in failsafe mode, the compressed spring is expanded and the hydraulic fluid, which is supplied to the actuator through the green line, is discharged from the actuator through the same line. The discharged hydraulic oil inside the green line is connected to a DCV.

In an open loop hydraulic system, illustrated at the top of Fig. 5.1, the hydraulic fluid enters the DCV and is discharged to the sea through a green line located under the DCV in the open loop schematic. For a closed loop system, the hydraulic oil discharged to the DCV through a return line is moved to topside through a return line and related umbilical.

5.2.2 Comparison

Open and closed hydraulic systems may be compared from different aspects; these are *operability and response time*, *reliability*, *logistics and installation* and *environmental issues*.

5.2.2.1 Operability and response time

The operability and response time of the valves illustrated in Fig. 5.1 are compared in open and closed loop systems. The assumption for the valves in the figure is that they are both FSC valves that open with hydraulic oil pressure and close through spring force. The opening time and characteristics of the actuated valves illustrated in Fig. 5.1 for both open and closed loop systems are similar while the hydraulic fluid moves inside the actuator cylinder and pushes the spring back to open the valve. However, the response and closing time of an actuated valve connected to a closed loop system in Fig. 5.1 is slower than the response and closing time of a valve connected to an open loop system. In both cases, a signal is sent to the DCV from the control room to close the connected actuated valve. As a result, the DCV changes its position to stop the hydraulic supply to the actuator and allow the hydraulic oil inside the actuator cylinder to be moved out into the hydraulic return line. In fact, there is a delay in closing a valve connected to a closed loop system due to back-pressure and pressure increase in the return line, as illustrated in Fig. 5.2. This delay does not exist for a valve connected to an open loop system as per Fig. 5.3.

A valve connected to a closed loop system has a delay in closing due to pressure build-up in the hydraulic return line. This build-up restricts the hydraulic fluid flow inside and makes closure of the valve and actuator more difficult. For operation in shallow water depths, there is no problem with restriction in the return line and no delay in the response time of the valve. But when it comes to deep-water field development, the spring cannot produce sufficient force to close the actuated valve and overcome the flow restrictions against the umbilical or tubing restriction in the return

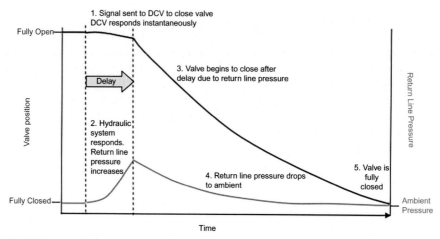

FIG. 5.2

Valve closure and pressure changes in a closed loop system.

Courtesy: URS Cooperation Limited.

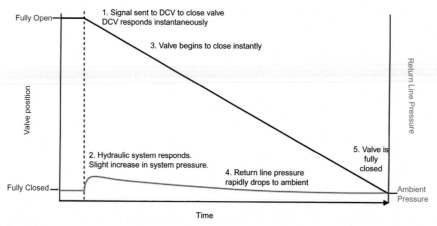

FIG. 5.3

Valve closure and pressure changes in an open loop system.

Courtesy: URS Cooperation Limited.

line. In that case, using one of the following solutions or a combination of both can solve the problem:

- Increasing the spring strength and torque in the actuator. Higher spring strength can be achieved by increasing the wire diameter of the spring.
- Increasing the return line and umbilical diameter so the tubing restriction is decreased.

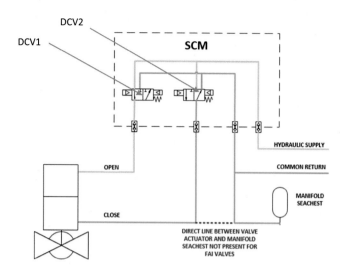

FIG. 5.4

Double-acting subsea actuator connected to a common return line

For a valve connected to an open loop system, the closing operation of the valve starts with a slight build-up of back-pressure in the system, as illustrated in Fig. 5.3; this does not last long since the oil is rapidly discharged into the marine environment.

The main question is what would be the response time and operability for fail-as-is double-acting actuators? Double-acting actuators work with hydraulic oil for both opening and closing. Fig. 5.4 illustrates a double-acting actuator connected to a common return line. The hydraulic supply passes through DCV1, enters the double-acting actuator valve and makes the valve open. While the actuator and the connected valve are opening, the hydraulic fluid is moving out from the other side of the actuator through the blue line. The hydraulic fluid moving out of the actuator through the blue line enters DCV2 and then flows to the common return line. But how does the valve get closed? A signal coming from the control room changes the position of both DCV1 and DCV2. DCV1 stops the hydraulic supply to the actuator through the orange line to open the actuator. The orange line, which is used to open the actuator, gets connected to the return line in DCV2. The changing position of DCV2 leads to the connection of the hydraulic supply line and fluid to the blue line, which closes the actuator. Thus after receiving the signal from the control room to change the positions of the DCVs, the hydraulic supply gets to the actuator through the blue line and closes the actuated valve. Conversely, the hydraulic fluid exits the hydraulic actuator on the open side and enters the common return line. Again, some sort of restriction exists for both opening and closing the actuated valve connected to the common return line, unlike double-acting actuators connected to an open loop hydraulic system. The restriction in the closed loop system could act against the opening torque or force produced inside the actuator, which reduces the safety factor of the actuator. The safety factor of the actuator is defined as the ratio of actuator torque or force to valve torque or force. Thus two approaches may be taken in dealing with double-acting actuators connected to a common return line in a closed loop system:

- Using a standard double-acting actuator with a reduced safety factor;
- Increasing the size of the double-acting actuator to achieve the required safety factor. Selecting the latter approach requires a new design and most likely a new qualification program to validate the design.

It should be kept in mind that the design and size of the remote operated vehicle (ROV) override could be subject to change due to the connection of the actuator to a common return line. ROV overrides are buckets installed on the actuators to provide possibility of the actuation independently from the hydraulic oil through manual ROV.

5.2.2.2 Reliability
Both open and closed loop systems are reliable and have been proven in many subsea fields. However, open loop hydraulic systems are simpler to maintain and it is easier to find and diagnose a problem in them if it occurs.

5.2.2.3 Logistics and installation
The main difference between an open loop and a closed loop system is that a closed loop system requires more umbilicals. In addition, actuators connected to a closed loop system may be bigger and bulkier than actuators connected to an open loop system. Therefore an open loop hydraulic system is a less costly option regarding the logistics and installation.

5.2.2.4 Environmental issues
As mentioned earlier in this chapter, an open loop hydraulic system is not eco-friendly due to the discharge of a large amount of oil into the marine atmosphere. Thus it is important to reduce the impact of environmental pollution if an open loop system is selected due to certain advantages such as lower cost, less required space, faster response time of the valves and actuators, etc. Thus if an open loop control system is selected for hydraulic subsea distribution, it is better to select water-based rather than oil-based hydraulic fluid to minimize the impact on the environment. Four water-based hydraulic fluid candidates with the least environment impact have been used in different projects across the world, in locations such as Africa, the Gulf of Mexico and Australia; these are proposed as follows:

- Castrol Transaqua HC10;
- Castrol Transaqua HT2;
- Niche Products Pelagic 100; and
- MacDermid Oceanic HW760R.

5.3 Case studies

5.3.1 Closed loop hydraulic system in a subsea project

This section reviews a case study of closed loop hydraulic distribution system selection that was implemented recently for a large subsea project including eleven manifolds. The subsea valves and actuators in this project were located at the depth of 2.3 km under the sea level.

Fig. 5.5 illustrates case #1: a subsea hydraulic closed loop circulation system for a subsea project. The schematic provides a brief description of how the hydraulic fluid moves inside the SCM and the actuated valves located on a subsea manifold. The hydraulic fluid is returned to the surface through the green lines, which are labeled "return line." The hydraulic fluid is then recirculated for usage inside the subsea actuators. The actuated valves in the schematic could have either spring-return or double-acting actuators. There is no hydraulic pressure supply to valve #1, which

Close loop system with circulation

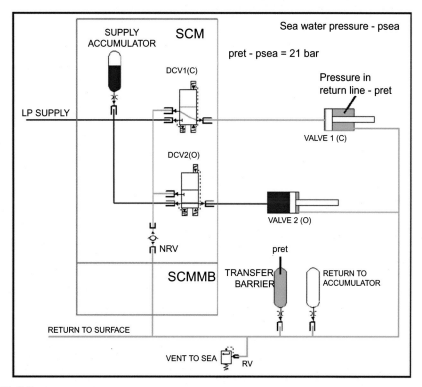

FIG. 5.5

Closed loop subsea hydraulic circulation system—case #1.

is in closed position. Conversely, valve #2 is in open position because the supply of hydraulic oil has passed through DCV #2 to move inside the actuator. The pressure of the oil supplied to an actuator at a subsea depth is equal to the hydraulic pressure at the surface where the HPU is located, plus the hydraulic head. The relevant formulas and calculations for determining hydraulic pressure at depth are described further in the case study. Both the supply and return hydraulic lines are equipped with accumulators to provide stable hydraulic flow rate without any pulsation. Assuming that both valves are actuated with an FSC actuator, the spring chambers of both valves are connected to a common return line.

The difference between the hydraulic pressure and seawater head is given equal to 21 bar. But how can the value of 21 bar be calculated? The hydraulic pressure in this case is equal to 345 bar or 5000 psi for valves and actuators installed at a depth of 2300 m. The seawater head is calculated in Formula (5.1).

Seawater head (SWH) calculation

$$SWH = G_a * \rho * D \qquad (5.1)$$

where SWH: seawater hydraulic head (Pa); G_a: gravity acceleration $= 9.81 \, \text{m/s}^2$; ρ: seawater density (kg/m^3); D: seawater depth (m).

Thus the seawater hydraulic head produced at a depth of 2300 m on subsea valves is equal to $232.6e^5$ Pa as calculated below:

$$SWHH = 9.81 \, \text{m/s}^2 * 1031 \, \text{kg/m}^3 * 2300 \, \text{m} = 23,262,453 \, \text{Pa} = 232.62 \, \text{bar}$$

Formula (5.1) is used to calculate the hydraulic oil head at a depth of 2300 m, considering that the hydraulic facilities located topside, i.e. the HPU and accumulators are located 20 m above sea level. The hydraulic oil type is MacDermid Oceanic HW760R which is a water-based hydraulic. The density of MacDermid Oceanic HW740R is placed in the formula for hydraulic head calculation.

$$\text{Hydraulic head} = 9.81 \, \text{m/s}^2 * 1114 \, \text{kg/m}^3 * (20 + 2300) \, \text{m} = 25,353,749 \, \text{Pa} = 253.54 \, \text{bar}$$

The difference between the hydraulic head and seawater head $= 253.54 - 232.62 = 20.92 \, \text{bar} \sim 21 \, \text{bar}$.

A relief valve (RV) is installed on the common return line to release overpressure into the sea. Thus it can be said that even using a closed loop system is not 100% environmentally friendly. The criteria for sizing the pressure relief valve are outside the scope of this chapter. However, in the author's experience in a couple of subsea projects, a set point of 60 bar was selected for the pressure relief valve in the common return line. The set pressure of the pressure safety valve should not exceed the maximum working pressure of the common return line.

Fig. 5.6 illustrates case #2: a closed loop subsea hydraulic circulation system. In this case, DCV #1 opens and allows flow of the hydraulic fluid to the actuator of valve #1, such that valve #1 opens. While valve #1 is moving to open position, the pressure drops at the supply accumulator and on the side of the actuated valve where the hydraulic supply is used to open the valve. The fluid inside volume #2 of the valve is moved to the common return line and creates back-pressure that acts

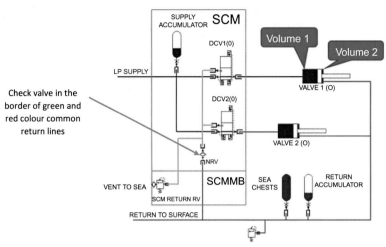

FIG. 5.6

Closed loop subsea hydraulic circulation system—case #2.

against the volume #1 hydraulic fluid for opening the actuator connected to valve #1. In this case, the pressure of the common return line fluid acts against the opening of the valve rather than its closing, unlike the schematic shown in Fig. 5.1. In fact, an area under the spring chambers of both actuators is connected to the common return line in this case.

An accumulator is installed on the common return line to prevent fluid flow pulsation and fluctuation. More precisely, this accumulator supports the system by absorbing some part of the hydraulic fluid in the common return line, which reduces the back-pressure of the fluid in the return line and thus aids the actuator in opening.

The common return line in Fig. 5.6 is connected to DCV #1 and 2 through a line that is red and green in case #2. The red part is separated from the green part through a check valve that is highlighted in the picture to prevent the possibility of returning the fluid in the green part to the red line.

Case #3, illustrated in Fig. 5.7, indicates valve #1 completely in open position and volume #1 completely pressurized with hydraulic fluid. Consequently, the fluid in volume #2 is discharged to the common return line and the hydraulic fluid, shown in green, returns to the (HPU). The return line pressure will be reduced and will ultimately stabilize.

5.3.2 Effect of a closed loop system on actuator design

A $7^{1/16''}$ slab gate valve with a linear fail safe closed hydraulic actuator is considered for the case study in this part. The pressure class of the valve is 517 bar equal to 7500 psi. The slab gate valve is installed on the branch line of a subsea manifold. The subsea manifold is installed at a water depth of 2000 m below sea level.

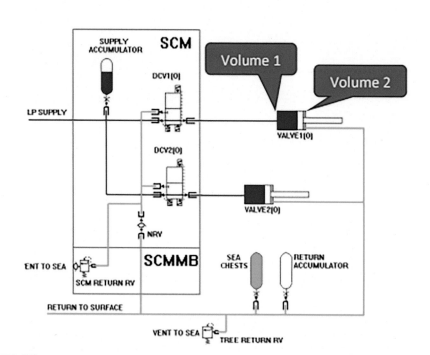

FIG. 5.7

Closed loop subsea hydraulic circulation system—case #3.

The valve is normally open to allow flow of fluid from the wellhead to the manifold. During normal operation when the valve and connected actuator are open, hydraulic oil with a pressure of 345 bar equal to 5000 psi pressurizes the hydraulic actuator cylinder to keep it open. If something were to go wrong inside the well connected to the subsea gate valve, a signal would close the valve through the spring force inside the actuator. The actuated valve is connected to a closed loop system as illustrated in Fig. 5.8. The bottom part of the actuator is connected to a common return line or sea chest through a component labeled "Hydraulic connection for common sea chest compensation system." The hydraulic supply oil enters the hydraulic actuator through item #74 and moves the piston or shaft down to open the valve. However, the presence of hydraulic fluid at the bottom of the actuator and common return line back-pressure resists against the hydraulic power for opening the valve and the connected linear actuator. On the other hand, stoppage of the hydraulic supply fluid into the actuator leads to the expansion of the spring to provide force or torque for closing the valve and actuator. The presence of the common return line oil back-pressure and the accumulated oil at the bottom of the actuator connected to the common return line assists the spring to close the valve and actuator. The spring chamber is located outside of the section illustrated in Fig. 5.8 on the top part of the illustrated actuator.

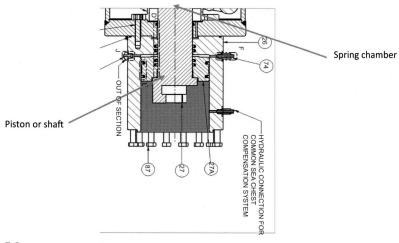

FIG. 5.8

$7^{1/16}''$ gate valve linear actuator.

The safety factor of the linear actuator as per the valve and actuator project specification should be at least 2. This means that the actuator should produce a thrust or force value at least 2 times that of the valve during the whole opening and closing operation time. The safety factor is calculated according to Formula (5.2).

Safety factor calculation

$$\text{Safety factor} = \frac{\text{Actuator torque}}{\text{Valve torque}} \qquad (5.2)$$

Different torque values are associated with valves, such as break to close (BTC), break to open (BTO), end to open (ETO), end to close (ETC) and maximum allowable stem torque (MAST), which are explained in more detail in Chapter 3. Table 5.1 illustrates the different torque values for opening and closing a $7^{1/16}''$ slab gate valve and connected linear actuator. The actuator torque should always be greater than the

Table 5.1 $7^{1/16}''$ gate valve and actuator torque values and safety factors.

	BTC (N)	BTO (N)	ETO (N)	ETC (N)	MAST (N)
$7^{1/16}''$ slab gate valve 517 bar	9000	25,000	8000	23,000	95,000
Hydraulic FSC linear actuator	70,000	60,000	45,000	55,000	Not applicable
Safety factor	7.78	2.4	5.62	2.39	Not applicable

valve torque to maintain the required safety factor. It should be noted that the torque values given in the table are approximate and are rounded to a whole number. All the values given in the table are in Newton (N) force.

The safety factors provided in the table indicate that the minimum safety factor of 2 is achieved by the forces created by the actuator compared to the connected valve.

MAST is an important parameter that depends on different factors, such as the material and diameter of the valve stem. Stem of valves in larger diameters and having higher mechanical strength have higher MAST. But why is MAST important? MAST indicates the maximum torque or force that can be applied to a valve stem without any damage. The MAST value for the valve in the case above is 95,000 N, which means that if the actuator were to provide a force higher than 95,000 N on the valve stem, the valve stem would be subject to damage. However, the maximum actuator force is 70,000 N, which is less than the MAST. Thus it can be concluded that the transmission of force from the actuator to the valve can be achieved without any damage to the valve stem.

What would be the effect on actuator torque values if the actuator were connected to a closed loop system with a common return line? As discussed earlier, a common return line acts against actuator opening, so both the BTO and BTC thrust values provided in Table 5.1 for a standard actuator would be reduced. In addition, the hydraulic oil in the common return line and the cavity at the bottom of the actuator illustrated in Fig. 5.8 assists the spring to close the valve, so the BTC and ETC thrust values provided in Table 5.1 for a standard actuator would be increased. The safety factor on BTC is 7.78 as per Table 5.1 which is much higher than the minimum acceptable safety factor of 2. Thus it is most likely that the effect of a common return line and closed loop system would not reduce the BTC of the actuator in such a manner that the BTC safety factor would go below 2. However, the safety factor for BTO as per the table is 2.4, which is not much higher than the minimum accepted safety factor of 2. This means that the BTO of the actuator could be reduced in such a manner that the safety factor for BTO would go below 2. Thus the aim of this section is to calculate the effect of a common return line on BTO and BTC.

The first step is to calculate the hydraulic supply pressure into the actuator at 2000 m sea depth. The hydraulic supply pressure is 345 bar equal to 5000 psi at the surface. The hydraulic pressure is reduced from 345 bar at a depth of 2000 m. The amount of reduction depends on the seawater and hydraulic head at a depth of 2000 m. Formula (5.1) is used to calculate both the seawater and hydraulic head at a depth of 2000 m (2 km).

Seawater head $(SWH) = 9.81 \, m/s^2 * 1031 \, kg/m^3 * 2000 \, m = 20{,}228{,}220 \ Pa = 202.28 \, bar$

Formula (5.1) is used to calculate the hydraulic head pressure at a depth of 2 km. It is assumed that the HPU installed on the topside facilities is 30 m above sea level.

Hydraulic head $(HH) = 9.81 \, m/s^2 * 1114 kg/m^3 * (30 + 2000) \, m = 22{,}184{,}530 Pa = 221.84 \, bar$

The difference between the hydraulic head and seawater head $= 221.84 - 202.28 = 19.56 \, bar.$

The difference between the hydraulic head and seawater head, calculated as equal to 19.56 bar, leads to a reduction of hydraulic pressure value at the depth of 2 km. Formula (5.3) is used to calculate the hydraulic pressure.

Hydraulic pressure at depth = hydraulic pressure at the surface
$\qquad$ − the difference between the hydraulic and seawater head

$$(5.3)$$

Using Formula (5.3) → $345 - 19.56 = 325.44$ bar $= 32.54 \frac{N}{mm^2}$.

The hydraulic force required to start opening a subsea valve at depth is calculated through Formula (5.4).

Required hydraulic force to start opening the actuator

$$F(\text{BTO}) = P@\text{depth} \times A \qquad (5.4)$$

where F: hydraulic force (BTO) (N); P: hydraulic pressure $(\frac{N}{mm^2})$; A: actuator cylinder area (mm^2).

$$\text{Using Formula (5.4)} \rightarrow A = \frac{F(\text{BTO})}{P} = \frac{60,000}{32.54} = 1843.88 \text{ mm}^2.$$

$$A = \pi r^2 \rightarrow r^2 = \frac{1843.88}{3.14} = 587.22 \rightarrow r = 24.23 \text{ mm} \rightarrow \text{cylinder diameter (d)} = 48.46 \text{ mm}.$$

The spring force for closing the valve and the BTC force are calculated according to Hooke's Law, given in Formula (5.5).

Hook's law

$$F = K * X \qquad (5.5)$$

where F = force (N); K = spring constant (N/M), assuming 700,000 N/M; X = spring extension (M).

Formula (5.5) → 70,000 (BTC) = 700,000 * X → X = 0.1 m = 10 cm = 100 mm.

The usage of Formula (5.5) shows that the spring is expanded 10 cm to generate BTC force and start closing the valve and the connected actuator. The next step is to show the effect of the common return line on both BTO and BTC forces. As discussed earlier, the pressure accumulated at the bottom of actuator acts against the opening of the actuator. Hydraulic analysis has been performed in the project to simulate the hydraulic pressure on the common return line. The result of the hydraulic analysis report shows that the hydraulic pressure in the cavity and common return line can reach a maximum of 70 bar. A pressure value of 70 bar in the cavity and common return line equal to $7 \frac{N}{mm^2}$ would thus be the worst-case scenario for opening the actuator. The maximum pressure in the common return line is largely dependent on the size of the common return line, such that enlarging it would reduce the pressure in the line. Assuming that the diameter of the area where hydraulic oil is accumulated at the bottom of actuator is equal to 50 mm, then it is possible to calculate the force in the cavity or bottom section of the actuator, parameter F_C, through Formula (5.4):

Maximum force created by hydraulic in common return line acts against opening of the actuator (F_C)

$$F_C = P * A = P * \pi r^2 = 7 * 3.14 * (50 * 50) = 13,738 \text{N}$$

Considering the effect of the common return line on cavity pressure, the new BTO_n can be calculated as follows:

$$BTO_n = 60,000 - 13,738 = 46,262 \, N$$

The torque of the valve is constant, so it is possible to calculate the safety factor of the actuator with the new BTO force as follows:

$$\text{Safety factor} = \frac{46,262}{25,000} = 1.85 < 2$$

Therefore the conclusion is either to accept the safety factor of 1.85 without any modification of the actuator, or to modify the actuator design in such a way that a higher BTO force is produced by the hydraulic supply force for opening. In the latter case, one solution is to increase the hydraulic oil supply cylinder diameter. The BTO of the modified actuator should be 50,000 N at minimum, after the reduction of 13,738 N, to maintain a safety factor of 2. Thus the new BTO for the modified actuator should be $50,000 + 13,738 = 63,738 \, N$.

$$\text{Using Formula (5.4)} \rightarrow A = \frac{F \, (BTO) \text{ modified}}{\text{Pressure of hydraulic at depth of 2 km}} = \frac{63,738}{32.54} = 1958.8 \, mm^2$$

$$A = \pi r^2 \rightarrow r^2 = \frac{1958.8}{3.14} = 623.8 \rightarrow r = 24.98 \, mm \rightarrow \text{cylinder diameter}$$
$$(d) = 49.96 \, mm \rightarrow \text{cylinder diameter increase} = 49.96 - 48.46 = 1.5 \, mm.$$

Therefore the cylinder diameter should be increased 1.5 mm to compensate for the hydraulic return line back-pressure. This is probably not a costly modification in itself. But it leads to qualification of the actuator, which imposes additional time and cost to the delivery of the actuator to the project.

Alternatively, the pressure of the hydraulic fluid helps close the valve and actuator, so it is possible to use a spring with a lower spring constant, i.e. a weaker spring compared to the standard design. The change of spring reduces the cost of the spring as well as the actuator. The new BTC force (BTC_n) is calculated as follows:

$$BTC_n = BTC - F_C \rightarrow BTC_n = 70,000 - 13,738 = 56,262 \sim 57,000 \, N$$

$$\text{Formula (5.5)} \rightarrow 57,000 \, (BTC_n) = K'*0.1 \rightarrow K' = 570,000 \, N/M.$$

Therefore the constant of the spring could be reduced from 700,000–570,000 N/M.

5.4 Conclusion

This chapter has briefly reviewed hydraulic systems for providing clean hydraulic fluid to subsea systems for the operation of valves and actuators. The advantage of a closed loop hydraulic system as the more environmentally friendly solution compared to an open loop system was discussed briefly. The recommendation in this chapter is to use a closed instead of an open loop hydraulic system in order to reduce hydraulic emission from the actuators to the marine atmosphere. The main outcome of using a closed loop hydraulic system is a cost increase due to the additional umbilical and piping or tubing system to return the hydraulic fluid to the surface after it has been used inside the actuators. The other challenge of using a closed loop subsea hydraulic system is the possibility of needing to modify the actuator design, which would necessitate new qualification programs. Two real cases of using closed loop hydraulic systems are discussed in this chapter. The first looks at the whole picture of using a closed loop system, including accumulators and pressure relief valves as well as additional umbilical and return lines. The second case is that of a $7^{1/16''}$ subsea slab gate valve in pressure class 517 with a linear spring return FSC actuator located on a manifold branch. It was concluded that the piston rod and cylinder diameter of the FSC linear actuator would need to be increased by some millimeters due to the hydraulic oil accumulation at the bottom of the actuator. The hydraulic pressure in a closed loop system helps to close the actuator and the spring force, so the spring constant and torque should be reduced as a result.

Questions and answers

1. Which sentence is correct about using closed loop hydraulic systems?
    A. Closed loop hydraulic systems are 100% environmentally friendly.
    B. Closed loop systems are less costly than open loop systems.
    C. Closed loop system usage could necessitate a change in the design of the actuators as well as an additional qualification program.
    D. Closed loop systems are not as reliable as open loop systems.

Answer: Option A is not correct; closed loop systems require hydraulic fluid, so they are not 100% environmentally friendly like all-electrical systems. In addition, safety relief valves installed on the return line of closed loop hydraulic distribution systems are used to release the overpressure hydraulic fluid into the sea. Option B is not correct, because closed loop systems contain more lines and umbilicals to transport the hydraulic fluid used inside the actuators to the surface, so closed loop systems are more expensive than open loop systems. Option C is correct, since using a closed loop hydraulic system can apply back-pressure or pressure restriction against opening or closing the actuator, so the design of the actuator could be modified, which could necessitate new qualification tests. Option D is not correct, since closed loop systems are just as reliable as open loop systems.

2. Which sentence is not correct?
    A. There is no need to select between an open and closed loop system for all-electrical subsea field development.
    B. Selection between a closed or open loop hydraulic distribution is an important decision for fields with electro-hydraulic power supply for subsea fields.

    C. Typically, water-based hydraulic oil, which has the least negative effect on the environment, is selected for open loop hydraulic systems.

    D. Actuators connected to a closed loop system always have shorter response time compared to actuators connected to an open loop system.

Answer: Option A is correct, because open and closed loop systems are hydraulic and do not pertain to all-electrical subsea systems. Option B is correct and, concerning the environmental issue, the proposal is to choose a closed loop hydraulic system. Option C is correct; an open loop hydraulic system is not an environmentally friendly choice, so if the choice is made, due to cost-saving or other benefits, to select an open loop hydraulic circuit, then water-based instead of oil-based hydraulic fluid should be used. Option D is wrong, since the connection of a closed loop system to a subsea actuator can cause some delay in the opening and/or closing time of the actuator due to the hydraulic back-pressure in the return line of a closed loop system.

3. Which sentences are not correct about the common return line?

    A. A common return line is a hydraulic line that transports the hydraulic oil used inside hydraulic actuators to the topside facilities.

    B. Back-pressure in the common return line can cause a delay in the response time of hydraulic actuators.

    C. There is no need to install a pressure relief valve and accumulator for a common return line.

    D. A common return line is typically connected to the spring side of a subsea actuator in a closed loop hydraulic system.

Answer: Option A is correct, since a common return line is a hydraulic line that transports the hydraulic fluid used inside hydraulic actuators. Option B is correct, because the connection of a common return line to the actuators could cause a delay in the operation and response time of the hydraulic actuators. Option C is not correct, because both a pressure relief valve and an accumulator are required to be installed on a common return line. The relief valve is installed to prevent an overpressure scenario in the line and an accumulator is required to prevent flow pulsation in the common return line. Option D is not correct, because the common return line could be connected to a spring chamber or hydraulic cylinder. In conclusion, both options C and D are incorrect.

4. Which sentence is correct about MAST?

    A. MAST could be less than the highest actuator force or torque.

    B. MAST can be reduced by increasing the valve stem diameter.

    C. MAST should be higher than all the produced force and torque of both the valve and actuator.

    D. Changing the stem material of a valve from 22Cr duplex to Inconel 718 reduces the MAST.

Answer: Option A is not correct, because having a MAST less than the highest actuator torque or force will cause valve stem damage. Option B is not correct, since increasing the stem diameter improves the resistance of the stem against the loads, so MAST is increased as well. Option C is correct; the MAST should be higher than all of the valve and actuator torque values to prevent any damage to the stem during valve operation. Option D is not correct, because Inconel 718 has higher mechanical strength compared to 22Cr duplex, so upgrading the stem material to Inconel 718 from 22Cr duplex would increase the MAST.

5. The linear actuator of a gate valve is connected to a common return line from opposite sides of a spring chamber. The common return line pressure acts against opening and helps with closing the valve and actuator. Which sentence is correct in this case?
   A. Increasing the size of the common return line leads to easier opening of the valve.
   B. The overall BTC force in this actuator is less than it would be for a standard actuator design.
   C. The overall BTO force in this actuator is higher than it would be in a standard actuator design.
   D. The safety factor of the actuator is the same in both standard design and the design connected to a closed loop hydraulic system and common return line.

Answer: Option A is correct, because increasing the size of the common return line results in less pressure in the common return line, so it is easier to open the valve against the common return line back-pressure. Option B is not correct, because although the actuator is connected to a common return line, the overall BTC force is due to spring force plus the pressure inside the common return line, so the overall BTC force is higher than it would be in a standard actuator design where closure is achieved solely due to spring force. Option C is not correct, since when the actuator is connected to a common return line, the overall opening force is equal to the overall force created during the standard actuator design for opening minus the hydraulic return line force, so the overall BTO force is less than it would be in a standard actuator design. Option D is not correct, because the safety factor of an actuator connected to a common return line is typically different from that of a standard actuator design.

References

[1] Y. Bai, Q. Bai, Subsea Engineering Handbook, first ed., Elsevier, Atlanta, GA, USA, 2012.
[2] K. Sotoodeh, The importance of maximum allowable stem torque in valves, Springer Nature Appl. Sci. 1 (2019) 433, https://doi.org/10.1007/s42452-019-0445-0.
[3] K. Sotoodeh, Actuator sizing and selection, Springer Nature Appl. Sci. 1 (2019) 1207, https://doi.org/10.1007/s42452-019-1248-z.
[4] URS Cooperation Limited, Environmental and Social Impact Assessment for Shah Deniz Stage II Project (Online), 2016 (accessed 6 December 2020).

Subsea actuator compensation system development for reducing emission

6.1 Introduction

Subsea development is moving toward deeper and more remote water depths, and subsea valves and actuators should be designed, manufactured and tested to prove the maintenance-free operation of this equipment in harsh and corrosive marine environments. The depth of actuator installation could be as deep as 3–4 km in some cases. At such depths, the seawater column applies a great amount of external pressure on both valves and actuators, which has a high tendency to implode them. As a rule of thumb, each 10 m of water depth produces 1 bar of pressure on the external surfaces of valves and actuators. Therefore if an actuator were installed at a water depth of 3500 m, an external water pressure of 350 bar would be applied to the actuator; the same pressure is applied to the external surface of subsea valves. The external pressure applied by the seawater to the valves and actuators should be compensated to prevent implosion. To compensate for this external seawater pressure, subsea valves are designed and manufactured to be thicker, based on higher pressure classes, according to American Petroleum Institute (API) standards such as API 6A or API 17D. As an example, the subsea ball valve illustrated in Fig. 6.1 is located at a depth of 3000 m. The valve internal fluid pressure at sea level is equal to 345 bar. The fluid pressure at a depth of 3000 m is equal to 300 bar, which is added to the fluid pressure at sea level. Thus a valve installed at a water depth of 3 km should be designed to withstand 645 bar pressure, which is almost 9352.5 psi (1 bar is almost 14.5 psi). It is noted that psi stands for pounds per square inch.

There is no 9352 psi as a standard pressure class in API 6A or API 17D standards. Typical API pressure classes are 2000, 3000, 5000, 10,000, 15,000, and 20,000 psi. A pressure class of 9352 psi is between 5000 and 10,000 psi pressure classes. A valve designed according to 10,000 psi due to seawater head is much heavier and thicker compared to a valve designed for installation at sea level with an internal pressure of 345 bar equal to 5000 psi. While selecting thicker valves to withstand the additional pressure is a logical countermeasure, it is not practical or cost-effective for actuators. The same load of 300 bar due to the seawater depth of 3 km applies to subsea actuators. If the actuator were designed and manufactured thicker, like the valve, it would be very heavy and impractical. Therefore the alternative approach, which is common

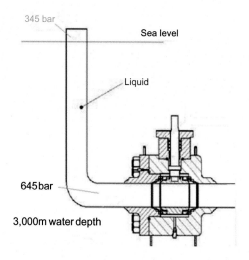

FIG. 6.1

A subsea ball valve at a depth of 3 km.

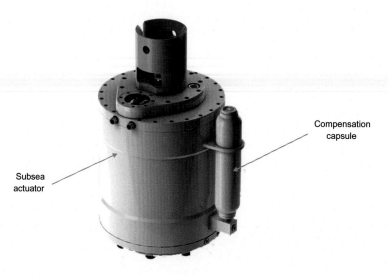

FIG. 6.2

A subsea actuator with a compensation capsule.

in the industry, is to connect a tank or capsule that is filled and pressurized with hydraulic oil to the casing of the subsea actuator to prevent it from imploding due to the seawater head pressure, as illustrated in Fig. 6.2. The connected capsule or tank is called a compensation system. The pressure inside the compensation capsule is almost the same as the seawater head pressure.

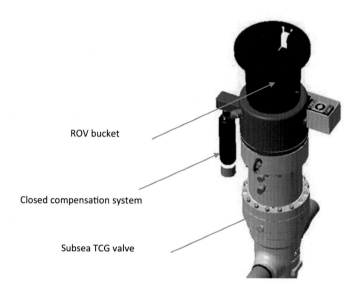

ROV bucket

Closed compensation system

Subsea TCG valve

FIG. 6.3

Closed compensation capsule on an ROV-operated gate valve close to the ROV bucket

In addition to actuators, subsea compensation systems can be installed on valve gear boxes and remote operated vehicles (ROVs). Fig. 6.3 illustrates a closed compensation system located close to the ROV bucket of a subsea through conduit gate (TCG) valve.

The benefits of using compensation systems are not limited to the prevention of component and facility collapse due to seawater head pressure. Compensation systems prevent the ingress of seawater into actuators and gear boxes and provide protection against pressure changes inside them due to solar radiation, operation of the gear box and actuator or subsea installation.

6.2 Types of compensation systems

There are two types of subsea compensation systems: closed and open. The main aim of this section is to introduce each type of compensation system, along with its advantages and disadvantages, and to select the one that is more environmentally friendly with less leakage to the environment.

6.2.1 Closed compensation systems

Figs. 6.2 and 6.3 illustrate a closed compensation system connected to an actuator and ROV bucket, respectively. A closed compensation system, as the name states, is a closed capsule containing hydraulic oil that normally does not have contact with the seawater, unlike an open compensation system. This means that a closed compensation system is the more environmentally friendly choice. Fig. 6.4 provides a detailed schematic of a closed compensation system, which in this case is a closed and pressurized capsule. There is a rubber bladder or diaphragm shown in black inside the closed compensation system. It is made of elastomeric material, such as nitrile rubber which is known as nitrile butadiene rubber (NBR) or Buna N. An elastomer is a type of polymer with high viscosity and elasticity; elastomeric polymer sealing can return to its original shape after stretching out to a large extent. The hydraulic oil, also called compensation fluid, is located above the diaphragm, and flows into the actuator from there. Seawater enters the compensation capsule without any direct contact with the hydraulic oil, because the diaphragm separates the seawater and hydraulic oil compensation fluid from each other.

Fig. 6.5 provides an overview of a capsule-shaped closed compensation system installed on a spring return actuator. The closed compensation system is filled with compensation hydraulic fluid, illustrated in orange, which is connected and distributed in the actuator's spring chamber. The compensation oil is normally different from the actuator supply fluid, which is shown in green in the figure. The schema on the right side of Fig. 6.5 illustrates the actuator in open position and shows the seawater at the bottom of the compensation system in blue color at the lowest level; the schema on the left illustrates the actuator in closed position and shows the seawater in blue color in the compensation system at the highest level.

The main advantage of a closed compensation system is that it is environmentally friendly, as the oil is not discharged into the seawater under normal operating conditions. The Health, Safety and Environmental (HSE) friendly nature of this system could be the main reason to choose a closed compensation over an open one. However, there are some disadvantages associated with this system, such as the difficulty of changing the compensation oil. There is no easy access to the capsule internals for this purpose. Another disadvantage is that a closed compensation system requires two relief valves, one inlet and one outlet, in order to balance the internal pressure with that of the seawater. If the pressure inside the compensation system increases and exceeds the seawater pressure by more than 5 bar, as an example, one of the pressure relief valves should release the excess pressure from the compensation system to the sea. Thus closed compensation system can release hydraulic compensation fluid to the sea in case of overpressurizing and it is not 100% environmentally friendly. On the other hand, if the pressure at the seawater depth exceeds the compensation system's internal pressure, the extra pressure in the seawater should be released to the compensation system through the other pressure relief valve to balance the two pressure values. A pressure relief valve is a type of safety valve used in the oil and gas industry to release extra liquid pressure to the environment. Making two holes in the compensation cavity to insert the pressure relief valves adds additional leakage paths to the compensation system, reducing its reliability. The other disadvantage of this system is related to the potential failure of the diaphragm, which leads to a mixing of the compensation oil and water, which in turn allows water to enter the actuator and cause corrosion. In addition, filling the closed compensation system with the hydraulic oil is difficult. Closed compensation systems are normally made of carbon steel.

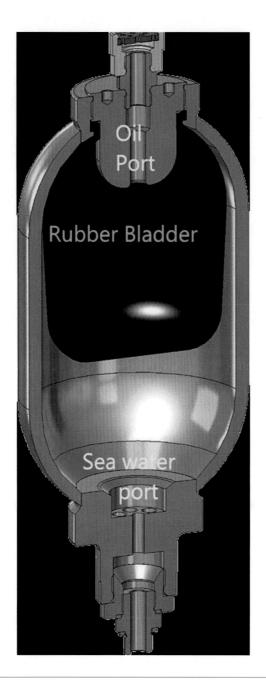

FIG. 6.4

Closed compensation system

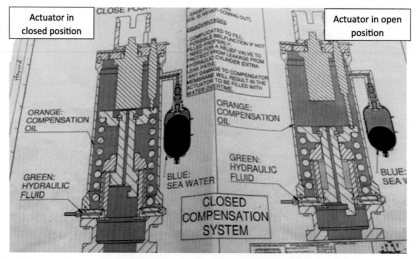

FIG. 6.5

Closed compensation system installed on a spring return actuator

Courtesy: Elsevier.

FIG. 6.6

Open compensation system

Courtesy: Petrol valve.

6.2.2 Open compensation systems

An open compensation system (see Fig. 6.6) is like a storage tank connected to the actuator casing; in this type of system, seawater directly enters the oil tank. The compensation oil in an open compensation system should have a higher density than the

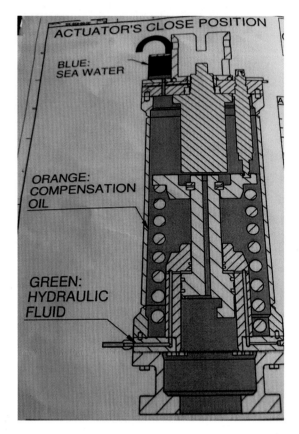

FIG. 6.7

Open compensation system on a spring return actuator

Courtesy: Elsevier.

seawater and not be mixable with it. These two requirements entail a separation between the oil and the seawater and prevent the seawater from entering the actuator.

Fig. 6.7 illustrates a spring return actuator in closed position compensated with an open tank located at the top left and filled in with seawater (highlighted in *blue*). The main advantages of open compensated systems are summarized as follows:

- There is no need for a pressure relief valve.
- It is easy to change the compensation oil.

Open compensation systems do have a major disadvantage, however: tilting the actuator leads to the release of compensation fluid into the sea, so this system is less environmentally friendly compared to a closed compensation system.

Open compensation systems should be made of a material compatible with seawater, such as 25Cr super duplex, or any material with a pitting resistance equal number (PREN) higher than 40, to avoid pitting corrosion. PREN is a measurement method used to evaluate the resistance of stainless steel against localized corrosion such as pitting; PREN depends on the material's chemical composition, specifically its percentage of chromium (Cr), molybdenum (Mo) and nitrogen (N), and can be calculated through Formula (6.1) below:

PREN calculation:

$$PREN = \% \, Cr + 3.3\% \, Mo + 16\% \, N$$

$$25 \, Cr \, super \, duplex \, PREN = 25 + 3.3 \times 3.5 + 16 \times 0.22 = 25 + 11.55 + 3.52 = 40.07 \quad (6.1)$$

Questions and answers

1. Which subsea components/facilities may have a compensation system?
   A. Valves and actuators
   B. Actuators and gear boxes
   C. Valves and subsea structures
   D. Subsea structures

Answer: According to the explanations provided in this chapter, using a compensation system to eliminate the effect of seawater head pressure is applicable to actuators, gear boxes and ROVs, so option B is correct.

2. Which sentence is completely correct about a closed compensation system?
   A. A closed compensation system does not have any chance of spillage to the environment.
   B. Filling a closed compensation system with hydraulic oil is easier than filling an open compensation system.
   C. A closed compensation system has less emission to the environment compared to an open compensation system.
   D. The hydraulic oil in a closed compensation system is in direct contact with the seawater.

Answer: Option A is not correct, because one of the relief valves installed on a closed compensation system releases the extra pressure of compensation oil to the seawater. Option B is not correct, since changing the compensation oil in a closed compensation system is a challenge due to the isolation of the compensation oil behind the nitrile rubber bladder. Option C is correct, since a closed compensation system is more environmentally friendly and generates less emission to the environment compared to an open compensation system. Option D is not correct, because the hydraulic oil in a closed compensation system is separated from the seawater through a bladder or diaphragm.

3. Which materials are typically selected for closed and open compensation systems?
   A. Carbon steel for a closed compensation system and a material with a high PREN for an open compensation system such as super duplex
   B. Austenitic stainless steel for both
   C. 6MO for both
   D. None of the above

Answer: Option A is correct.

Further reading

American Petroleum Institute (API) 6A, Specification for Wellhead and Tree Equipment, twenty-first ed., American Petroleum Institute (API), Washington, DC, USA, 2018.

American Petroleum Institute (API) 17D, Design and Operation of Subsea Production Systems, Subsea Wellhead and Tree Equipment, second ed., American Petroleum Institute (API), Washington, DC, USA, 2011.

K. Sotoodeh, Subsea Valves and Actuators for the Oil and Gas Industry, first ed., Elsevier, USA, 2021.

HIPPS for fugitive emission reduction

7.1 Flare system

Natural gas is produced as a byproduct of oil during the production process. The cost of installing a pipeline to transmit the produced gas to the market is high in many cases, so the remaining choices are either to inject the gas into the reservoir to improve oil recovery or burn it in a flare. Other factors may contribute to making the decision to flare the gas, such as lack of infrastructure, lack of a commercial market to buy the produced gas or a long distance between the field and an existing commercial market.

Put briefly, a flare is a part of a safety system that burns overpressurized hydrocarbons and releases them into the environment. A flare system is typically filled in with gas when the system becomes overpressurized, and a pressure relief or safety valve opens to release the overpressurized gas or liquid to the flare system. All of the released gases and liquids are routed through a large pipe called a piping header to a vertical standing flare. The gases in the flare system are burned in the flare stack.

Fig. 7.1 illustrates a pressure vessel labeled "Protected System." The overpressure fluid in the vessel opens the pressure relief valve (PRV) installed on the top of the vessel, and the fluid moves toward the flare system. The pressure source in the main header and subheaders to the flare apply back-pressure to the PRV, which is outside the scope of this chapter.

7.2 Flaring and fugitive emission

Flare systems are elevated to keep the open flame away from ground level and reduce the effects of heat, smoke and noise. Although burning hydrocarbons in a flare, as illustrated in Fig. 7.2, rather than releasing them into the atmosphere without burning them, is a more environmentally friendly solution, a great amount of greenhouse gas is released into the environment through flaring. Some statistics show that more than 140 billion cubic meters of natural gas are flared annually. The flaring of natural gas worldwide in the oil and gas industry produces 400 million tonnes of greenhouse gas emissions. The main greenhouse gases that are released to the environment through flaring are methane (CH_4) and carbon dioxide (CO_2). In some cases, the gas may contain hydrogen sulfide (H_2S) which is a very toxic compound. Approximately one-fourth of the flare gas and greenhouse emission comes from offshore topside

Prevention of Actuator Emissions in the Oil and Gas Industry. https://doi.org/10.1016/B978-0-323-91928-9.00007-4

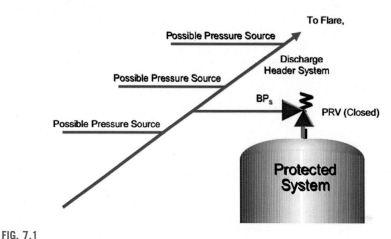

FIG. 7.1

Flaring overpressure fluid inside a pressure vessel.

FIG. 7.2

Burning of hydrocarbons in a flare tip.

Courtesy: Shutterstock

platforms. Thus the discussion in this chapter covers refineries, petrochemical plants and onshore and topside oil field developments. Flaring is not applicable to the subsea oil and gas industry. Therefore the purpose of this chapter is to focus on using a high integrity pressure protection system (HIPPS) to reduce flare capacity and thus flare emission to the environment. In some cases, a refinery or petrochemical plant may require expansion and an increase in the capability of the pressure relief valves and connected flare lines. Implementation of HIPPS is a solution that can be considered in all of the above-mentioned cases in order to reduce flare capacity.

7.3 **HIPPS for flare reduction**

High integrity pressure protection systems (HIPPS) have been used in the oil and gas industry for many years to prevent overpressurizing scenarios. Applying HIPPS has several benefits, such as reducing the size and capacity of the flare system, which results in less fugitive emission to the environment, downrating the piping located after the HIPPS system in some cases, especially in subsea systems, and fast isolation of the system in the event of detecting a high-pressure scenario to protect personnel and equipment and reduce capital cost. The focus of this chapter is on the first advantage: reducing the size and capacity of the flare system, thus reducing fugitive emission. But before addressing the capacity of HIPPS to reduce fugitive emission, it is important to know the HIPPS components, their function and significance.

According to NORSOK P-001, 'Process design' and P-100 'Process system' standards, HIPPS systems are typically used to reduce the instance relief rates for overpressurizing protection; thus using HIPPS facilitates reduction in the dimensions of flare systems and reduces flare emission.

HIPPS is the outcome of evolution in the design of mechanical and electronic safety devices used in process systems. In fact, HIPPS is a part of a safety instrumented system (SIS) designed to prevent overpressurizing by shutting of the overpressure piping system. SIS requirements are typically defined according to International Electrotechnical Commission (IEC) standards 61,508 and 61,511. These two international standards provide requirements for SIS specification, design, installation, operation and maintenance to ensure that the system is situated and maintained in a safe state. Both standards are generic and do not address HIPPS specifically. All HIPPS components, which will be discussed later, should follow IEC requirements to verify and validate their compliance with functional safety requirements. The sensors and safety instrumentations should be in accordance with IEC 61511. SIS always contains three elements: sensors for measurement, a logic solver for making decisions and a final element that performs a task. One of the challenges related to the implementation of HIPPS in the topside offshore oil and gas industry is that no standard exists to address HIPPS directly. Traditionally, American Petroleum Institute (API) standard 14C has been used for the implementation of safety systems in the offshore sector of the oil and gas industry; API 17O now directly addresses HIPPS in the subsea sector of oil and gas industry, which is outside the scope of this chapter.

HIPPS is also known as "high integrity pressure piping system" because it deals with and is located on piping or pipeline systems. API 521, the standard for pressure relieving and de-pressuring systems, defines HIPPS as an arrangement of instruments, final control elements (e.g. valves, switches, etc.) and logic solvers (see Fig. 7.3).

Pressure transmitters (see Fig. 7.4) monitor pipeline/piping pressure against a predefined limit and send signals to the logic solver where an appropriate decision is taken based on the nature of the signal. The type of signal transmitted from the pressure sensors to the logic solver (see Fig. 7.5) has different voting logic, such

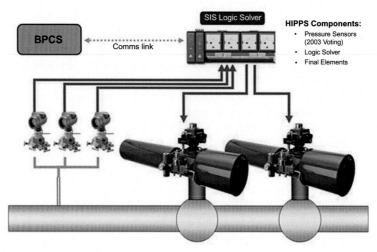

FIG. 7.3

HIPPS architecture.

Courtesy: Emerson Automation.

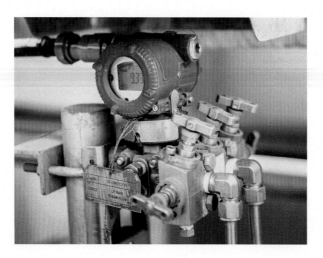

FIG. 7.4

Pressure transmitter.

Courtesy: Shutterstock.

as 2 out of 3 (2oo3). One of the key parameters in the design of SIS is the voting arrangement or architecture. The main concept behind using a voting system is redundancy—using extra components to tolerate the failure of one component. Redundancy consideration can reduce the risk of failure and improve the system's safety integrity level, which is discussed in detail later in this chapter. There is a

FIG. 7.5

Logic solver.

IMI Critical Engineering.

different SIS architecture for each voting arrangement. There are two numbers used for voting; the first is the number of devices that should vote for a trip to occur. Trip is a type of action performed by some systems, usually a SIS, to put an industrial process to a safe condition. The second is the total number of installed devices or components. Therefore in the case of a 2oo3 vote, two of the devices out of three agree on the shut-down of the system, so this will happen. 2oo2 voting indicates that both devices (two out of two) must agree to shut down the system. The logic solver output in the form of an electrical signal results in an action taken by the final element, such as shutting down the process system by closing a valve.

7.4 HIPPS justification

Until 1996, ASME section VIII, division 2 required usage of pressure relief valves on every single pressure vessel. Subsequently, however, both API 521 and ASME Section VIII, divisions 1 and 2 began to allow the usage of a SIS system such as HIPPS to act as the preliminary overpressure protection barrier on condition that the SIS system provides better protection and higher reliability. The reliability of HIPPS is evaluated through quantitative risk assessment or analysis (QRA) and proof of the safety integrity level (SIL) of the HIPPS system by means of a safety analysis report (SAR). HIPPS as a part of SIS should maintain the required safety and reliability level over the life-cycle of the plant. Safety and reliability life-cycle is the part

Table 7.1 SIL levels according to IEC 61508.

SIL	PFD	Risk reduction factor
SIL 1	0.1–0.01	10–100
SIL 2	0.01–0.001	100–1000
SIL 3	0.001–0.0001	1000–10,000
SIL 4[a]	0.0001–0.00001	10,000–100,000

[a]*Not applicable in the process industry.*

of a management system that is used to ensure a functional safety system. SIL is introduced and recommended in IEC 61508 and 61,511 standards to measure reliability and probability of failure in the safety system, such as HIPPS. The relationship between SIL and probability of failure on demand is provided in Table 7.1, based on IEC 61508. Many companies have chosen SIL 3 for HIPPS applications; SIL 3 does not allow more than 0.001 or 0.1% chance of failure. Maintaining SIL 3 for HIPPS was a challenge before in the industry, but it is achievable now.

Overall risk analysis to justify using HIPPS instead of a conventional flare system based on ASME section VIII and API 521 should contain the following steps:

Step 1: Is the pressure vessel used in air, water or steam? If yes, the using HIPPS is not permissible.

Step 2: Do legal authorities ask for conventional design with a flare system? If yes, then HIPPS is not a correct solution.

Step 3: Conducting a hazard analysis of an overpressure scenario and selecting the required SIL: Each scenario of overpressurization where the usage of conventional pressure safety valve and flare lines are not sufficient should be evaluated and proven. This evaluation could be done by a team including process, safety and operation engineers as a minimum. The second important decision in step 3 is to evaluate the required SIL for HIPPS.

Step 4: Can HIPPS meet the required company risk guideline(s)? If the amount of risk tolerated by the company guideline(s) is lower than HIPPS risk and reliability, then selecting HIPPS is not recommended.

Steps 5 and 6: Development of detail design and safety requirement for HIPPS: Typically, a specification for HIPPS according to relevant standards such as IEC 61508 and IEC 61511 should be developed. Some calculations should be performed to determine whether the required SIL for the HIPPS system is achievable.

Step 7: Is usage of HIPPS safe or safer than using a PRV? According to different experiences, pressure relief valves that are sized properly and maintained well during operation can provide safety integrity level 2 or 3. It has been experienced that many pressure relief valves have maintained SIL 3 historically. Thus considering SIL 3 for HIPPS and proving it through documents such as a safety analysis report (SAR) could prove that HIPPS is either safer or as safe as PRVs. Maintaining HIPPS level 3 was a challenge in the past. But many valve and actuator suppliers and manufacturers of pressure transmitters can maintain SIL 3 for HIPPS nowadays.

Diagnosis is a fundamental feature of safety instrumented systems, because using diagnostics can change the classification and probability of a large portion of dangerous failures (λ_D) into dangerous detected failures (λ_{DD}). This change can increase the safe failure fraction (SFF) and SIL, which results in improving the reliability of the system. SFF is calculated through the ratio of the average probabilities of safe (λ_S) plus dangerous detected failures (λ_{DD}) to the safe (λ_S) and dangerous failures probabilities (λ_D) using Formula (7.1):

SFF calculation:

$$\text{SFF} = \left(\frac{\lambda_S + \lambda_{DD}}{\lambda_S + \lambda_D} \right) \tag{7.1}$$

where λ_S: probability of safe failure; λ_D: probability of dangerous failure; λ_{DD}: probability of dangerous detected failure.

There is a correlation between SFF and SIL based on the number of fault tolerances in the system, as shown in Table 7.2. As per the table, the tolerances in the system for the created faults could be zero, one or two.

All of the components that are used in a HIPPS system, such as sensors (pressure transmitters) and logic solvers, have an internal diagnostic feature. However, HIPPS actuated valves do not have a diagnostic feature. Thus it is proposed to apply a test, such as partial stroke test, in order to increase SIL, safety and reliability and reduce downtime. A partial stroke test is an important safety step to assure that on/off valves function safely during operation. The advantage of partial stroke testing is that the valve is not fully closed, so there is no need to stop or interrupt production during the test. This type of testing is used for HIPPS and for emergency shut down valves. It should be noted that emergency shut down valves are considered as the main category of safety critical valves. A partial stroke test can be conducted a couple of times per year to detect possible failure modes of the actuated valve during some percentage of valve opening, i.e. 'partial stroke.' The operator should check the operability of the valve and actuator during the partial stroke test and make sure that the communication between the valve and actuator is functioning correctly.

Step 8: Removing the excess flare load: The last step is to consider HIPPS during safety relief valve and flare sizing.

Fig. 7.6 illustrates a flow chart for determining whether HIPPS can be implemented instead of a conventional flare system as per the eight steps mentioned above.

Table 7.2 Correlation between SFF and SIL.

Safe failure fraction	Hardware fault tolerance – Type A		
	0	1	2
SFF < 60%	SIL 1	SIL 2	SIL 3
60% < SFF < 90%	SIL 2	SIL 3	SIL 4
90% < SFF < 99%	SIL 3	SIL 4	SIL 4
SFF ≥ 99%	SIL 3	SIL 4	SIL 4

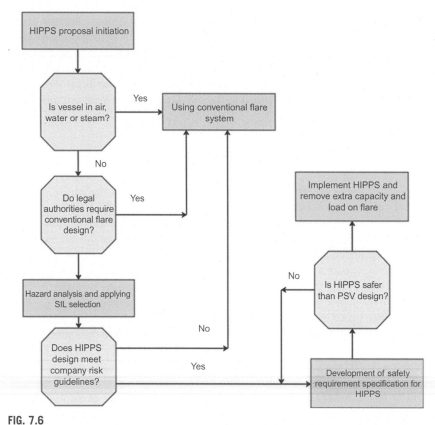

FIG. 7.6

HIPPS implementation flow chart.

7.5 HIPPS and flare reduction case study

The main aim of this section is to explain and demonstrate how using HIPPS in the onshore and offshore topside sectors of the oil and gas industry leads to a reduction in flare size and load. Fig. 7.7 illustrates a simple process flow diagram for an oil and gas production unit from the wellhead to the separator where three phases of gas, water and oil are separated without the implementation of HIPPS.

As shown in Fig. 7.7, the produced hydrocarbon entering from the wing valve located on the wellhead passes through a shutdown valve (SDV). The next step, a choke valve, is a pressure regulating or pressure control valve that reduces fluid pressure significantly. As illustrated in the figure, the operating pressure of the fluid is reduced from 476 to 40 bar because of the choke valve. The fluid then enters the separator, a type of pressure vessel, where the liquids are separated from the gas. The liquids exit the separator from the bottom (low point), while the gases leave the separator from the top (high point). There is a pressure safety valve (PSV) installed on

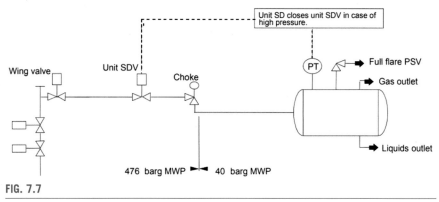

FIG. 7.7

Hydrocarbon production and separation without HIPPS implementation.

Courtesy: AIS.

the high point or top of the separator to release any overpressure. In fact, the PSV is considered the final overpressure protection device in the separator. The first overpressure protection system in the separator is the emergency shut down valve. A pressure transmitter (PT) is installed on the top of the separator; it measures the pressure inside the separator. In the event that it senses and detects any overpressure scenario in the separator, the pressure transmitter transmits a signal to the shut down valve (SDV) to close the production line. Stoppage of the production line causes pressure reduction in the separator. The second safeguard for overpressure protection in the separator is the choke valve. Closure of the choke valve causes production shut-down as well as pressure reduction in the separator. If the SDV does not close (failure #1) and the choke valve does not close (failure #2), then overpressure in the separator is discharged through the PSV located on the top of the separator. In this case, the PSV should be sized for full flow of the well and a flare piping system is necessary.

Fig. 7.8 illustrates a simple process flow diagram of an oil and gas production unit from the wellhead to a separator where three phases of gas, water and oil are separated with HIPPS implementation. In this case, if both the shut down valve and choke valve fail to close the line, two HIPPS valves can close it in very short period—such as 0.2 s. Thus a small PSV can be installed on the top of the separator and the flare line could be smaller in size than in the previous case.

It should be noted that an axial valve, as illustrated in Fig. 7.9, is the best choice of valve for HIPPS, given its proven reliability compared to other types of valves such as gate and ball valves. Axial valves have an internal structure like that of axial check valves, but unlike axial check valves, axial valves have a stem that transfers the actuator force to the valve internals to open and close the valve. There are a variety of reasons that make axial valves the most reliable choice for HIPPS application, such as reliability and fast operation—particularly closing.

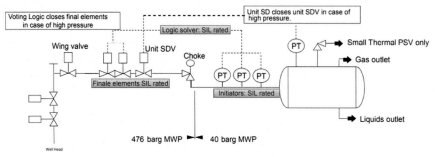

FIG. 7.8

Hydrocarbon production and separation with HIPPS implementation.

Courtesy: AIS.

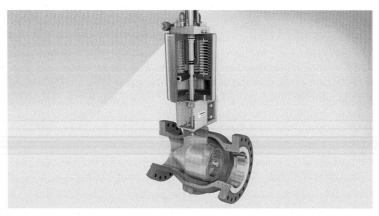

FIG. 7.9

Actuated axial valve for HIPPS application.

Courtesy: Mokveld.

Questions and answers

1. Which sentence is not correct about flaring?
   A. Flaring is used for hydrocarbon gases that are difficult to sell.
   B. Flaring is a more environmentally friendly solution compared to releasing hydrocarbon into the atmosphere without burning it.
   C. Flare lines are typically located after pressure relief valves.
   D. Flaring the gas into the environment does not release a high amount of greenhouse gases.

Answer: Options A, B and C are correct. Option D is not correct, since flaring releases a considerable amount of greenhouse gases into the environment.

2. Which sentence is correct about HIPPS?
- **A.** A HIPPS includes an actuated valve, pressure transmitters and a logic solver.
- **B.** Using a HIPPS is always more reliable than using a conventional flare and safety relief valve.
- **C.** Maintaining SIL 3 for HIPPS is rarely possible.
- **D.** IEC 61508 and 61511 specifically address HIPPS.

Answer: Option A is correct. Option B is not always correct, since the reliability of HIPPS and safety relief valves are largely dependent on their safety integrity level. As an example, if SIL 3 is maintained for both HIPPS and a safety relief valve, then HIPPS is as reliable as the safety relief valve. Option C is not correct, since maintaining SIL 3 for HIPPS was a challenge before but not now. Option D is not correct, because IEC 61508 and 61511 do not specifically address HIPPS.

3. One of the challenges associated with HIPPS arises from SIL 3 implementation. How can SIL 3 be maintained in HIPPS?
- **A.** Applying partial stroke testing for the HIPPS actuated valve
- **B.** Selection of axial valves
- **C.** Applying self-diagnostics for the logic solver and sensors
- **D.** All options are correct

Answer: Option D is correct, since all the options are correct.

Further reading

American Petroleum Institute (API) RP 14C, Analysis, Design Installation and Testing of Safety Systems for Offshore Production Facilities, eighth ed., API, Washington, DC, 2018.

American Petroleum Institute (API) 17O, Standard for Subsea High Integrity Pressure Protection System (HIPPS), second ed., API, Washington, DC, 2014.

American Petroleum Institute (API) 521, Pressure Relieving and De-Pressuring Systems, seventh ed., API, Washington, DC, 2020.

American Society of Mechanical Engineers (ASME), Design and Fabrication of Pressure Vessels. Boiler and Pressure Vessel Code. ASME Section VIII Div. 02. New York, NY, 2012.

International Electrotechnical Commission (IEC), IEC 61508, Functional Safety of Electrical/Electronic/Programmable Electronic Safety-Related Systems, second ed., International Electrotechnical Commission (IEC), Geneva, Switzerland, 2010.

International Electrotechnical Commission (IEC), IEC 61511, Functional Safety—Safety Instrumented System for the Process Industry Sector, second ed., International Electrotechnical Commission (IEC), Geneva, Switzerland, 2016.

NORSOK P-001, Process Design, fifth ed., NORSOK P-001, Lysaker, Norway, 2006.

NORSOK P-100, Process Systems, second ed., NORSOK P-100, Lysaker, Norway, 2001.

E.M. Marszal, K.J. Mitchell, Justifying the use of high integrity pressure protection systems (HIPPS), in: ASME Pressure Vessel & Piping Conference, 2004.

D. Picard, Fugitive emission from oil and natural gas activities, J. Good Pract. Guid. Uncertain. Manag. Natl. Greenh. Gas Invent. (1999) 103–127.

K. Sotoodeh, Safety integrity level in valves, J. Fail. Anal. Prevent. 19 (2019) 832–837, https://doi.org/10.1007/s11668-019-00666-2.

K. Sotoodeh, Subsea Valves and Actuator for the Oil and Gas Industry, first ed., Elsevier, USA, 2021.

A.E. Summers, High integrity protection systems for flare load mitigation, in: NC: Instrumentation, Systems, and Automation Society, Houston, Texas, October 20–23, 2003, p. 2003.

C. Thoegersen, Reduce fugitive emissions with a HIPPS solution, Valve World Mag. (2013) 33–37.

Subject Index

Note: Page numbers followed by *f* indicate figures and *t* indicate tables.

Author Index

Printed in the United States
by Baker & Taylor Publisher Services